Who Pays for Clean Water?

Westview Replica Editions

This book is a Westview Replica Edition. The concept of
Replica Editions is a response to the crisis in academic and
informational publishing. Library budgets for books have been
severely curtailed; economic pressures on the university presses
and the few private publishing companies primarily interested in
scholarly manuscripts have severely limited the capacity of the
industry to properly serve the academic and research communities.
Many manuscripts dealing with important subjects, often representing the highest level of scholarship, are today not econom-
ically viable publishing projects. Or, if they are accepted for
publication, they are often subject to lead times ranging from
one to three years. Scholars are understandably frustrated when
they realize that their first-class research cannot be published
within a reasonable time frame, if at all.

Westview Replica Editions are our practical solution to the
problem. The concept is simple. We accept a manuscript in camera-
ready form and move it immediately into the production process.
The responsibility for textual and copy editing lies with the
author or sponsoring organization. If necessary we will advise
the author on proper preparation of footnotes and bibliography.
We prefer that the manuscript be typed according to our specifications, though it may be acceptable as typed for a disserta-
tion or prepared in some other clearly organized and readable
way. The end result is a book produced by lithography and bound
in hard covers. Initial edition sizes range from 500 to 800
copies, and a number of recent Replicas are already in second
printings. We include among Westview Replica Editions only works
of outstanding scholarly quality or of great informational value,
and we will continue to exercise our usual editorial standards
and quality control.

Who Pays for Clean Water?
The Distribution of Water Pollution Control Costs

Elizabeth E. Lake, William M. Hanneman, and Sharon M. Oster

Who will pay for water pollution control? How great will the burden be for various socioeconomic groups? Will the distribution of costs be equitable? In answering these questions, the authors examine changes in public-sector budgets resulting from the Water Pollution Control Act, as well as the industrial response to the act. They conclude that clean water can be financed in two ways--public agencies can pay, generating funds through taxes, fees, or the reduction of other public services, or industries can recover their expenditures through increased prices to consumers. Finally, they evaluate the burdens that will be placed on different segments of the population as a consequence of each method.

Elizabeth E. Lake is vice president of Urban Systems Research and Engineering, Inc. William M. Hanneman is assistant professor of agricultural and resource economics at the University of California, Berkeley. Sharon M. Oster is associate professor of economics at Yale University.

Other Urban Systems Research Reports

Housing for the Elderly: An Evaluation of the Effectiveness of Congregate Residences, Irene K. Malozemoff, John G. Anderson, and Lidia V. Rosenbaum

The Definition and Measurement of Poverty, Volume 1: A Review, Sharon M. Oster, Elizabeth E. Lake, and Conchita Gene Oksman

The Definition and Measurement of Poverty, Volume 2: Annotated Bibliography, Sharon M. Oster, Elizabeth E. Lake, and Conchita Gene Oksman

Medicaid Eligibility: Problems and Solutions, Marilyn P. Rymer, Conchita Gene Oksman, Lawrence C. Bailis, and David T. Ellwood

Who Pays for Clean Water?
The Distribution of Water Pollution Control Costs

An Urban Systems Research Report
by Elizabeth E. Lake,
William M. Hanneman,
and Sharon M. Oster

Routledge
Taylor & Francis Group

LONDON AND NEW YORK

First published 1979 by Westview Press

Published 2019 by Routledge
52 Vanderbilt Avenue, New York, NY 10017
2 Park Square, Milton Park, Abingdon, Oxon OX14 4RN

*Routledge is an imprint of the Taylor & Francis Group,
an informa business*

Library of Congress Catalog Card Number: **79-5152**

ISBN 13: 978-0-367-21344-2 (hbk)
ISBN 13: 978-0-367-21625-2 (pbk)

Acknowledgments

The research on which this report is based was performed for the National Commission on Water Quality under contract WQ5AC001. The authors are grateful for their support and guidance.

The authors would also like to thank Ms. Carol Cerf for her work as technical editor on the book, and Dr. James F. Hudson for his data processing and data management on the project.

Contents

xi

Exhibits

Who Pays for Clean Water?

I. Introduction

Many local governments and private industries will
install new treatment facilities in response to the 1972
Amendments to the Water Pollution Control Act, P.L. 92-500.
In the final analysis, individual Americans will pay the costs
of these facilities in many ways: through higher taxes,
reduced public services, and increased prices for the goods
consumed. This report examines how these costs are likely
to be distributed among households of differing social and
economic circumstances. We try to answer three related
questions: Who will pay for water pollution control? How
great will the burden be for different socioeconomic groups?
Will the distribution of costs be equitable?

Our answer is in two parts. On one hand, public
agencies building treatment facilities can pay the costs
through taxes, charges, or the reduction of other public
services. On the other hand, industries may be expected to
pay for treatment through increased prices. Each method of
financing will impose different burdens on the various
segments of our population.

The first part of this study examines changes in public
sector budgets (financing, revenues, expenditures) resulting
from the Water Pollution Control Act. Given the costs of
treatment required by the Act, we predict how those costs
will be paid. Knowing the total costs and the methods used
to finance them, we calculate the impact on subgroups of the
population.

The second part of the study examines the industrial
response to the Act. Industry finances the costs of mandatory
water pollution control investment partly through price
increases. The consumers of the higher priced products
feel a loss related to the size of the price increases, so,
to trace the impact of the industrial treatment requirements,

we examine how consumption of particular products is distributed among the subgroups of the population. The net cost to a subgroup is the sum of the municipal and industrial components.

It should be noted that the burdens estimated here include only the effects of increased prices, taxes, and reduced public services. Other burdens, such as losses in GNP due to unemployment, reduced economic growth, loss of corporate profit due to inability to pass costs on, or increases in interest rates and land acquisition costs, are excluded from this analysis.

HISTORY OF WATER POLLUTION CONTROL LEGISLATION

Dissatisfaction with existing water pollution control legislation led to the passage of the Water Pollution Control Act Amendments of 1972, P.L. 92-500.* The 1972 Amendments evolved from the substantial history of federal involvement in water pollution control. The new Act changed the emphasis of water pollution control and increased federal financial assistance for this purpose.

The first modern federal legislation dealing with water pollution control is the Refuse Act of 1899. This Act forbade the discharge of wastes into navigable waterways without a permit from the Army Corps of Engineers, but municipal discharges were exempt. The Act remained largely unenforced until 1970, when Secretary Hickel revived it to serve as the legal basis for prosecuting individual dischargers.

Aside from two minor pieces of legislation at the beginning of the century, recent federal legislative activity in this field began with the Water Pollution Control Act of 1948 (P.L. 62-115). This Act gave the federal government responsibility for surveys, research and investigation; however, the main responsibility for control remained with the states. Although the Act authorized loans for the construction of treatment facilities, no funds were ever actually appropriated.

The next major piece of federal legislation was the Water Pollution Control Act Amendments of 1956 (P.L. 84-660). The main features of this law include the provision of federal grants for the construction of treatment facilities and the establishment of procedures for prosecuting

---

*A.V. Kneese, Charles L. Schultze, Pollution Prices and Public Policy, Brookings Institute, Washington, D.C., 1975.

individual dischargers. These two features were dominant
in federal water pollution control legislation until 1972.
Under this Act, local governments could obtain federal grants
to cover up to 30% of the estimated reasonable cost, or
$520,000, whichever was smaller. The applicants had to make
provisions for the proper and efficient operation and
maintenance of the facility under construction, and they had
to demonstrate conformity with state plans. Fifty percent
of the'appropriated funds were to be allocated among states
on the basis of a population/income formula, and at least
50% had to go to communities with populations of 250,000 and
less. The first targets of the enforcement procedures were
interstate polluters, but the amendments passed in 1961
strengthened the law to include all dischargers into
navigable waters. The 1961 amendments also increased the
appropriations for wastewater treatment related construction
activity.

The Water Quality Act of 1965 was the next major
landmark in the history of federal water pollution control
legislation.* This Act increased the federal government's
enforcement powers, and made each state's water quality
criteria, implementation and enforcement plans subject to
federal review. At this time, the Federal Water Pollution
Control Authority (FWPCA) was created in the Department of
Health, Education and Welfare, but was later moved to the
Department of the Interior.

The 1966 Amendments, the Clean Water Restoration Act,
changed the grant allocation formula for individual projects
(although the distribution-among-states formula remained the
same). The federal share varied from 30-50 percent,
depending upon state contributions and actions with regards
to enforceable water quality standards.

The 1968 Amendments, the Water Quality Improvement Act,
provided for contracts of up to 30 years to pay for the
federal share of the construction costs. They could be used
towards the repayment of municipal bonds. In addition,
one-time grants were awarded for improving the operation of
existing facilities.

The 1970 Amendments (P.L. 91-224) initiated
demonstration projects for the Great Lakes and added control
measures for air pollution, hazardous materials, sewage from

---

*Meta Systems, Inc., Alternative Methods of Financing
Municipal Waste Treatment Facilities, prepared for EPA,
Spring 1974.

3

vessels, and mine water pollution.

In 1970, President Nixon issued Executive Order 11574 (based on the Refuse Act of 1899), which established a permit system for all point source dischargers. At this time the FWPCA was moved from the Department of the Interior into the newly created Environmental Protection Agency.*

In sum, the three most significant features of water pollution control legislation prior to the current law (P.L. 92-500) were (1) construction grant subsidies, (2) water quality standards, (3) a permit system for individual point source dischargers. Because of the continuing deterioration in the quality of the nation's waters, and because of incompatibilities between the permit system, the National Environmental Policy Act (NEPA), and the FWPCA authorities, the legislative program came to be considered as a failure. It was replaced by the Water Pollution Control Act Amendments of 1972, P.L. 92-500.

THE PROVISIONS OF P.L. 92-500

For the purpose of investigating the incidence of the municipal and industrial treatment costs of the Act, P.L. 92-500 has two features of major importance. These are:

- the increased emphasis on the permit system with litigation procedures against municipal and industrial dischargers who violate the system; and

- a substantial increase in and new rules for the federal construction grants program.

These features are described in Titles V and II of the Act, as well as in applicable EPA regulations.** +

---

*Environmental Policy Division, Congressional Research Service, Library of Congress, Congress and the Nation's Environment, Environmental and Natural Resources Affairs of the 92nd Congress, 1973, Washington, D.C.

**Public Law 92-500, 92nd Congress, S.2770, October 18, 1972.

+Federal Register, Vol. 39, No. 29, Part III, EPA, Water Pollution Control, Construction Grants for Waste Treatment Works.

The Permit System

All point source discharges must have a permit to dispose of their liquid wastes in the waterways of the nation. The permits, issued for specified time periods, embody the effluent limitations objectives of the Act. These are:

- for industries "best practicable" water pollution control technology by 1977, and "best available" technology by 1983;

- for publicly owned treatment plants, secondary treatment by 1977, and best practicable technology (BPT) by 1983;

- for some new types of industries, "best available" technology (BAT), and where possible, the discharge of no pollutants;

- more stringent treatment requirements in cases of conflict with water quality standards;

- "zero discharge" for both municipalities and industries by 1985.

Compliance with the treatment requirements of the permits is ensured through a system of self-monitoring and reporting to the EPA, spot checks by the state, and, when necessary, full scale investigation by EPA. Violators are prosecuted by the state or EPA, with penalties of heavy fines (on the order of $10,000 per day) to shutdown.

Industries using publicly owned treatment facilities do not need a federal permit, although municipalities may establish their own pre-treatment regulations for accepting effluents, and require permits in that process.

In addition to meeting local requirements, "significant"* industries discharging incompatible wastes to Publicly Owned Treatment Works are also subject to national pretreatment standards.

The Construction Grant Program

Federal grants for the construction of publicly owned facilities may cover 75% of the EPA approved construction

---

*A significant industry has been defined as one discharging 30,000 gallons per day or more.

costs. These costs may include facilities planning and the preparation of construction drawings and specifications as well as actual construction costs. In most cases, the states also contribute towards the costs of construction.

Each state is responsible for establishing a system of priorities for the allocation of federal funds, in terms of both amounts and timing. Initially, the scope of a construction project will be determined by the applicant and municipality. This may be revised in terms of the state priority schedule, and must be approved by EPA. Other grant requirements are the establishment of sewer ordinances to ensure proper treatment plant operation, and the development of "user charge" and "industrial cost-recovery" schemes.

Grant assistance under the Act is contingent upon schemes which ensure that industrial users of the treatment facilities will repay their proportional share of the federal grants, and upon user charge schemes allocating total operational, maintenance and replacement costs among all users or classes of users in proportion to their contribution to the total waste load.

The conditions established by the law for industrial cost recovery require signed letters of intent from significant industrial users (using 10% or greater of design capacity) agreeing to proportional repayment, and specifying the period of intended use. Proportionality must be based at least on effluent volumes but could include waste concentration characteristics as well. Payments from all industrial users may be spread over the life of the plant, up to a maximum of 30 years, and must be made at least annually. These payments are interest free. Of the recovered amounts, 50% is returned to the federal government. Of the remaining 50%, 80% and accrued interest must be allocated to the expansion and replacement of the treatment facility, with the remainder being used as the municipality sees fit. In fact, the remainder may be spent on expenditures not related to the treatment facility.

The development of user charge schemes is largely up to the municipality within the constraints of EPA regulations. User charge schemes are to allocate operations and maintenance (O&M) costs between different users or classes of users in proportion to their contribution to the total treatment works loading. Factors such as strength, volume, and delivery flow rate are used to ensure a proportional distribution of costs. The user charge schemes must be revised periodically to reflect changes in costs, and they must generate sufficient revenue to offset all O&M costs.

Related Legislation

In addition to P.L. 92-500 there are several other
federal legislative initiatives which affect the costs of
water pollution control. On the municipal side, two other
programs provide funds for wastewater management facilities.
The HUD Community Development Block Grants can be used for
this purpose, and the Economic Development Administration
still provides funds for wastewater management infrastructure
in eligible EDA Development Districts. Neither program is
explicitly considered in this report. For industrial water
pollution control, certain provisions of the federal tax
code affect the costs of new equipment. These are considered
only insofar as they ultimately are reflected in price
changes. State economic development programs which make
publicly assisted financing available to industries are
similarly omitted from the analysis.

ORGANIZATION OF THE REPORT

In addition to this Introduction, this report includes
four major chapters. Chapter 2 below presents some background
concepts used throughout this report. The concepts and
measures of income, income distribution and incidence are
detailed, and data on existing conditions are presented.

Chapters 3 and 4 treat, respectively, municipal
expenditures and industrial price increases. The background,
methodology, data, and results are presented in each. The
final chapter combines the incidence patterns, and assesses
the sensitivity of the results to changes in the assumptions
of the data and methodology.

7

II. Background Considerations

Several concepts, assumptions, and considerations are basic to the analysis and results of this report. Those touching on both the public sector and the industrial impacts are described in this chapter. Possible approaches to ascertaining equity are first discussed. Measures of income are then reviewed, followed by a discussion of methods of measuring vertical equity. The final section of this chapter provides a series of data on existing income distribution in the United States.

EQUITY

In order to proceed in this analysis, it is first necessary to define what is meant by an equitable or socially acceptable outcome. In the conventional literature of welfare economics, a "socially acceptable" outcome is the notion that government policy should involve equal sacrifices (or yield equal benefits) to individuals or households in similar circumstances. This concept is generally referred to as "horizontal equity."

But equal sacrifices of what? In the usual literature an equal "welfare" (or utility) sacrifice is required. Welfare in turn is generally related to economic capacity (or income) through a set of income-utility functions. Sacrifice is, then, related to income. It should be noted at this point that income is not the only measure of economic capacity; in colonial times, for example, property was a better measure. At present, consumption and net worth are possible alternative measures.

The argument over whether income is an appropriate measure of one's economic condition is an old one, and is most frequently articulated in tax literature. Thomas Hobbes*

---

*Thomas Hobbes, Leviathan, Part 2 (New York: Dutton, 1924), esp. Chapter 30.

argued, for example, that the appropriate base for taxation was consumption, not income--that people should be taxed on the basis of what they "took, out of the pot," not what they put into it. Since Hobbes' time, the debate has become somewhat more complex, but the issue remains essentially the same: Is affluence best measured by income, wealth, or consumption?

There are several problems with using annual income figures to identify economic condition. In any one survey year, the observed low-income group will include some people who are earning below their normal level, as the result of negative windfalls.* Moreover, low income people in particular tend to underreport income.** The Office of Business Economics of the U.S. Department of Commerce corrects to some extent for this underreporting using field survey data, but these adjustments are clearly problematic.

Consumption has frequently been offered as an alternative poverty definition. It is argued that consumption expenditures are a better proxy for permanent income than current income. The choice between consumption and income is far from a trivial one. Margaret Reid estimated that in 1960 for large U.S. cities, consumers with incomes under $1,000 were spending $224 for every $100 of income received; the average expenditure for the two families in Washington with incomes under $1,000 was $5,404.+

Wealth has become an increasingly popular alternative to income as a poverty index.++ The primary argument here is one of equity. In particular, it is argued that income from assets (wealth) is different in kind than income from labor; people who possess assets wealth have a kind of security not available to individuals at comparable income levels who have no wealth. A strict income definition ignores this distinction.

Because of difficulties in acquiring and analyzing distributional data on either consumption or net worth, this

---

*M. Friedman, A Theory of the Consumption Function, (New Jersey:Princeton University Press, 1957).

**Margaret Reid, "Poverty--Defining the Problem." Statement in the Hearings before the Subcommittee on the War on Poverty, U.S. House of Representatives, Apr. 23-28, 1964: p. 1427-1438.

+Margaret Reid, op.cit.

++Lester Thurow, "Wealth Taxes," in National Tax Journal, 1971, and The Impact of Taxes in the American Economy.

study used income as the basic index of economic capacity and welfare. Our underlying normative judgment is that individuals or households with equal incomes should be treated equally.

In the welfare economics literature, it is also generally acknowledged that people with different incomes should bear different--or progressive--burdens; that is, those with higher incomes should bear a higher proportion of the costs. This principle of progressive burden distribution is referred to as "vertical equity." The question of how progressive the burden distribution should be, however, is less clear. Thus, in this report we have used several different specifications for the ideal degree of progressivity in evaluating the Water Pollution Control Act, including the Gini index and comparisons with the existing tax structure.

In this study, we examined not only the way in which pollution abatement costs varied with income, but also the variance of these costs with a number of other population attributes. This analysis was conducted in two ways. First, we directly examined the variance of these costs with several of the more important horizontal descriptors, including city size, government structure of the community, and region of the country. The relevant horizontal cells are summarized in Exhibit II-1. Secondly, we traced through the effects of the income-cost relationship on individuals with certain attributes which are themselves correlated with income; race is a horizontal descriptor that was treated in this fashion.

Exhibit II-1

Cells Defined for Horizontal Descriptors

1. GEOGRAPHIC REGIONS

    Northeast   North Central
    South   West
                 Non-contiguous Areas

2. RACE

    Black   White and Others

3. CITY SIZE

    Population: Less than 5,000
                5,000-9,999
                10,000-24,999
                25,000-99,999
                100,000+

4. AGE OF HOUSEHOLD HEAD(S)

    Labor Force Age: 18-65   Retirement Age: 65+

11

Two criteria were used in the selection of horizontal descriptors. First, only population characteristics expected to be of interest to policy makers were considered. Thus, the distribution of costs by region and city size was considered; the distribution of costs by marital status was not. Secondly, we concentrated on those population characteristics which we expected to be related to costs, either directly, or indirectly through their link with income. A summary table of the methods used to trace the distribution of costs by socioeconomic class is provided in Exhibit II-2.

At this point, it should be noted that the analysis described here is limited to the distributional impacts of direct cost increases incurred through tax increases to support public treatment facilities, and price increases to cover the costs of industrial treatment. Other effects, such as reduced economic growth, regional unemployment caused by industrial plant closings and relocation, and interest rate increases, are not considered. The stimulative effects of the P.L. 92-500 public works program are also excluded from the analysis.

EXHIBIT II-2

Summary Statement of Methods Used to Trace
Distribution of Costs by Socioeconomic Class

1. Distribution of Costs by Race

   Derived From: (1) distribution by income
                    (2) distribution by city size
                    (3) distribution by region

2. Distribution of Costs by Age

   Derived From: (1) distribution by income
                    (2) distribution by region

3. Distribution by City-Size

   Directly calculated

4. Distribution by Region

   (a) direct effect calculated
   (b) additional effects due to the city size-regional
       relationship inferred

MEASURES OF INCOME

A major objective of this study is to determine the extent to which the burden of the costs of the water pollution abatement program falls on the rich and poor groups of the

population. Some definition of the relative incomes of
individuals is clearly required. One approach, of course,
would be to use the total money income of individuals as an
index of their relative affluence. The use of money income as
an index, however, is not without problems. In particular,
there are a number of non-monetary components of income* which
should be viewed as part of total income. Since this non-
monetary income is unevenly distributed among subgroups of the
population, the use of money income alone as a base or index
from which incidence is evaluated may bias the burden
estimates.

In the light of the problems discussed above in this
report, we have used a total adjusted income. Following
Musgrave,** we have defined total adjusted income as Census
income (adjusted for underreporting) plus transfer payments,
pre-tax retained corporate profits, employer wage supplements,
inputed rent on owner-occupied housing, insurance interest
and other capital gains. Musgrave provides data on the
distribution of national adjusted family income for 1968.
For years after 1968, we stretched Musgrave's income brackets
to reflect anticipated growth in income. Musgrave's income
groups and the stretching this study used for 1975 calculations
are shown in Exhibit II-3.

In order to estimate the total adjusted income for
geographical subdivisions, it was necessary to transform the
aggregate money income figures, as well as the money data on
states, geographic subdivisions, and city types into adjusted
income figures. This was done by applying the current
Musgrave estimates of the aggregate relationship between
money income and total income to the area income distributions.
The implicit assumption in this calculation is that the
distribution of income adjustments is uniform across time,
areas of the country, and city types. Proportionality
figures used in this procedure are those given in Exhibit II-3.

---

*An example of non-monetary income is food grown by
farmers for home-consumption.

**Richard A. Musgrave, Karl E. Case, and Herman Leonard,
"The Distribution of Fiscal Burdens and Benefits," Public
Finance Quarterly, July 1974.

```
        Exhibit II-3

  Income Cells Used for Households
          1968 & 1975


                    Total Income

Income   Money Factor   1968          1975
Group    Income

 (1)     $0-2,000       $0-4,000      $0-6,700
 (2)     $2,000-4,000   $4,000-5,700  $6,700-9,500
 (3)     $4,000-6,000   $5,700-7,900  $9,500-13,100
 (4)     $6,000-8,000   $7,900-10,400 $13,100-17,200
 (5)     $8,000-10,000  $10,400-12,500 $17,200-20,700

 (6)     $10,000-15,000 $12,500-17,500 $20,700-29,000
 (7)     $15,000-20,000 $17,500-22,600 $29,000-37,500
 (8)     $20,000-30,000 $22,600-35,500 $37,500-58,900
 (9)     $30,000-50,000 $35,000-29,000 $58,900-152,700
(10)     $50,000+       $92,000+      $152,700+

SOURCE:  1968:  Musgrave, Case & Leonard, The Distribution
         of Fiscal Burdens and Benefits, HIER Discussion
         Paper 319, September 1973.

         1975:  Musgrave categories stretched so that new
         brackets contain same relative number of families
         as the old brackets.
```

MEASURES OF VERTICAL EQUITY

Once the distribution of burdens among individuals from the water pollution abatement program is identified, one must still determine whether that distributional pattern is "good" or "bad." Unfortunately, there is little consensus on the ideal degree of progressivity or vertical equity. Therefore, in this report, we use two different measures of vertical equity: the Gini Index (and its graphic analogue--the Lorenz Curve) and comparisons with the present tax structure. In this section, each of these techniques is reviewed, and the yardstick or "ideal" degree of progressivity implicit in each is outlined.

The Lorenz curve is a commonly used method for presenting data on changes in the income distribution. The percentage of total income is graphed on the vertical axis, the percentage of the population--ranked from poorest to richest--along the horizontal. (See Exhibit II-4.) A perfectly equal distribution of income is then represented by the diagonal of the square; the

14

Exhibit II-4:   The Lorenz   Curve and Gini Coefficient

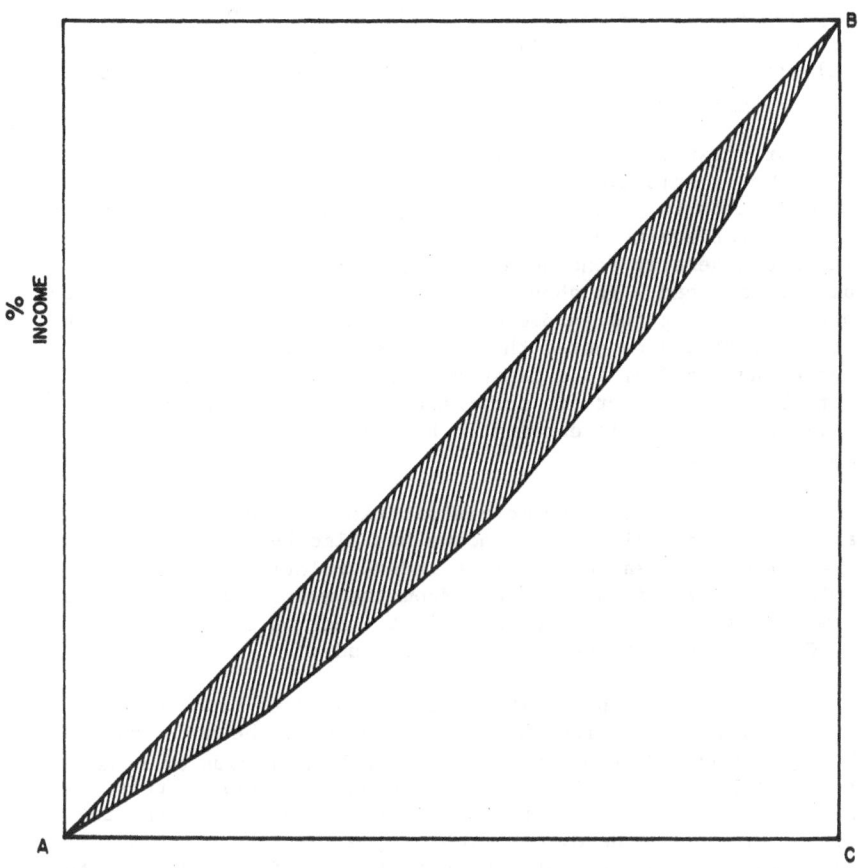

actual--unequal-- distribution may be given by the curve under
the diagonal as in Exhibit II-4. The hatched area is a measure
of inequality. The ratio of this area to triangle ABC is a
measure of the degree of inequality. It varies from 0 (perfect
equality) to 1 (perfect inequality). The Gini index is commonly
used by the Census Bureau as their summary measure of income
distribution.*

----

*A simple formula for calculating the Gini is:

$$G = 1 - \sum_{i-1}^{k} (f_{i+1} - f_i)(Y_i + Y_{i+1})$$

Where k = # of income brackets; i = income bracket;
f = cumulative distribution of family units; and
y = cumulative income distribution.

The value of the Gini Coefficient is extremely sensitive to the definition of brackets; therefore, the Gini coefficients calculated on the basis of census income brackets are not comparable with the Gini indices calculated for Musgrave's ten income brackets. This will not present a problem for this analysis, since changes in the Gini coefficient will be measured for the same ten income groups.

Our first method for evaluating the distributional effects of the Water Pollution Control Act was to identify the extent and direction in which the Act changed the Gini. The underlying criteria for equity used here is that programs which lower the Gini are generally considered "good;" conversely, programs which raise it are bad. In short, at least within the relevant range, more equality in the income distribution is deemed better than less. It should be noted that Sheshinski* has demonstrated that using the Gini as a measure of the "goodness" of particular distributional effects is consistent with the rather strong egalitarian principle enumerated by Rawls,** that is, the minimax strategy.

The slope of the income distribution provided by the variance of natural logs of income has also been considered as an alternative measure. In this index, relative weights are applied to deviations in income from the mean. However, since in most cases the variance index is a function of the Gini, only the Gini index is calculated in this study.

The second technique which we have used to probe the desirability of the distributional effects of the Act compares the cost distribution with the existing distribution of taxes. It should be noted that the yardstick which underlies this comparison is susceptible to two rather different interpretations. On the one hand, one might argue that the existing structure of tax rates reflects society's preferences as to the optimal degree of redistribution. In this case, the Act is "bad" if it leads to a cost distribution either more or less regressive than the current system. Alternatively, one might argue that the tax system reflects a minimum acceptable level of redistribution; in this case, only those cost distributions less progressive than the tax system are undesirable. Therefore, the distribution of the water pollution control cost burden was compared with the distribution of the personal income tax, the total (federal, state and local) tax burden, the property tax, and the user charge burden.

_____

*Eytan Sheshinski, "Relation Between a Social Welfare Function and the Gini Index of Income Inequality," Journal of Economic Theory, 1972, pp. 98-100.

**John Rawls, A Theory of Justice, Belknap Press of Harvard University, Cambridge, 1961.

Income Data

In order to calculate vertical incidence, data were needed on the shape of the income distribution for the nation, states, regions of the country, and various city types for the period 1970-1985.

Current money income for the U.S. and geographic subdivisions was taken from the Census; this was transformed into adjusted income estimates using the Musgrave proportionality constants discussed earlier in this chapter. These data are summarized in Exhibits II-5 through II-9.

In this report, we have calculated the financial burden of water pollution control for families as opposed to households, because the distributional data used has been tabulated for families only. Although the distribution of family income has not been formally compared with the distribution of household income, one may assume that the latter is slightly more unequal.* If this assumption is true, then the per family burden estimates calculated in this report are slightly overstated for the top and bottom income groups, and are slightly understated for the middle income groups.

Exhibit II-6 describes the regional distribution of income in the U.S. Poverty in the South is indicated by the relatively large number of families in the lower income groups, and small number of families in the higher income groups. The median income group for the South is the fourth; for the Northeast, it is the sixth; and for the North Central area, the West and the nation as a whole, it is the fifth.

The disparity between the distribution of income for smaller (under 10,000) and larger centers is striking, with the distribution being skewed toward the lower income groups in the smaller centers. This is presented in Exhibit II-7. For the smaller centers, income group #4 is the median group; for the larger centers it is income group #6.

----

*A number of studies have indicated a trend toward greater equality in family incomes, and toward greater inequality in individual incomes. See, Edward Budd, "Postwar Changes in the Size Distribution of Income in the U.S." American Economic Review (1970); T. Paul Schultz, Long Term Change in Personal Income Distribution: Theoretical Approaches, Evidence and Explanations, Santa Monica: Rand Corporation, 1972; John A. Brittain, The Payroll Tax for Social Security, Washington: Brookings Institution, 1972.

17

Exhibit II-5

The Distribution of Families by Income Groups (1975)

Income Group	% of Families	# of Families (millions)
1	12.3%	7.97
2	7.0%	4.54
3	9.8%	6.35
4	12.0%	7.78
5	14.3%	9.27
6	24.5%	15.87
7	7.8%	5.05
8	8.5%	5.51
9	3.0%	1.94
10	0.8%	.52
TOTAL	100.0%	64.80

SOURCE: % of families by income group was calculated from the 1970 Census of Population, Social and Economic Characteristics, United States Summary.

The total number of families for 1975 was obtained from the Census Population Projections, Series E.

Exhibit II-6

The Distribution of Families by Income Group
for Geographical Regions

Income Group	United States	North- east	North Central	South	West
1	12.3%	9.1%	10.4%	17.6%	10.5%
2	7.0%	5.6%	6.1%	9.2%	6.6%
3	9.8%	8.7%	8.8%	12.1%	9.1%
4	12.0%	11.5%	11.6%	13.0%	11.4%
5	14.3%	14.7%	15.1%	13.5%	14.0%
6	24.5%	26.5%	26.9%	20.2%	25.7%
7	7.8%	9.1%	8.4%	5.6%	8.8%
8	8.5%	9.9%	9.0%	6.1%	9.6%
9	3.0%	3.8%	2.0%	2.2%	3.5%
10	0.8%	1.0%	0.7%	0.6%	0.8%
Total	100%	100%	100%	100%	100%
% of U.S.	100%	24.6%	28.3%	30.2%	16.9%

SOURCE: Calculated from the 1970 Census of Population, op.cit.

Exhibit II-7

The Distribution of Families by Income for Selected City Sizes

Income Group	Total U.S.	Cities Over 10,000	Cities Under 10,000
1	12.3%	10.0%	16.6%
2	7.0%	6.1%	8.8%
3	9.8%	8.7%	11.9%
4	12.0%	11.2%	*13.5%
5	14.3%	13.8%	14.6%
6	24.5%	*26.3%	21.1%
7	7.8%	9.2%	5.2%
8	8.5%	10.0%	5.6%
9	3.0%	3.7%	1.8%
10	0.8%	1.0%	0.4%
Total	100%	100%	100%
% of U.S.	100%	66%	34%

SOURCE: Calculated from the 1970 Census of Population, op.cit.

*Median

```
┌─────────────────────────────────────────────────────────────┐
│                        Exhibit II-8                          │
│       The Cumulative Distribution of Income and Population    │
│  Income                                                       │
│  Group      Cumulative Income*       Cumulative Population**  │
│    1              4.0%                      12.3%             │
│    2              8.3%                      19.3%             │
│    3             14.8%                      29.1%             │
│    4             24.8%                      41.1%             │
│    5             37.5%                      55.4%             │
│    6             63.3%                      79.9%             │
│    7             77.0%                      87.7%             │
│    8             85.0%                      96.2%             │
│    9             92.0%                      99.2%             │
│   10            100.0%                     100.0%             │
│     *Calculated from Musgrave, Case & Leonard, op.cit.       │
│    **Calculated from Exhibit II-5                            │
└─────────────────────────────────────────────────────────────┘
```

Comparisons between distribution of income for particular subgroups of society are relevant for our purposes: equal per capita costs represent heavier burdens for lower incomes. Indeed, as will be shown later, the burden of water pollution control costs is the most regressive in the South and in the smaller population centers.

With regards to the distribution of income for the future, we have assumed that no major changes will occur over the time period considered. As income and population grow over time, we assume that the relative proportions of income and number of families remains constant. In other words, 12.3% of families continue to fall into income category #1, 7% into category #2, and so on. As income grows, the bracket limits of the categories are stretched, according to the methodology developed by Musgrave.* Growth in total income reflects both increases in national economic capacity and population growth. The income and population projections used in this study for stretching the bracket limits were the projections used in other Commission work. Income projections were developed by the Wharton Macroeconomic Model, while population forecasts were based on the Census series E forecasts. It was assumed that income and population growth would affect each income group equally, and thus the proportion of families in any income group would remain constant over time.

This assumption is not unreasonable in the light of recent trends. Although from 1930 to about 1947, there was signifi-

_____

*Musgrave, Case and Leonard, op.cit.

Exhibit II-9

Lorenz Curve: The Distribution of Income Without Water
Pollution Control Expenditures

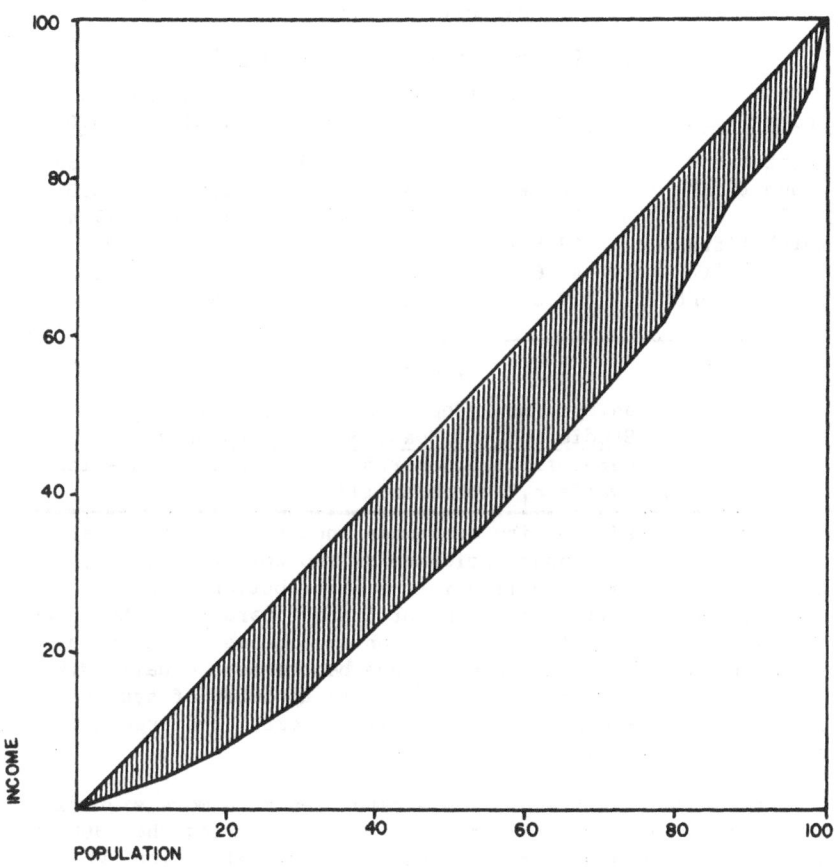

The diagram is based on Exhibit II-8.

The corresponding Gini value is 0.27.

cant movement toward greater equality in family incomes, that
trend has slowed considerably since 1950. Some indications of
the changes in the distribution of income are presented in
Exhibit II-10.

Exhibit II-10

Changes in the Distribution of Income: 1950-1970

Income Class	Percent of Population in Each Income Class				
	1950	1955	1960	1965	1970
Lowest fifth	4.5%	4.8%	4.9%	5.3%	5.5%
Second fifth	12.0%	12.2%	12.0%	12.1%	12.0%
Middle fifth	17.4%	17.7%	17.6%	17.7%	17.4%
Fourth fifth	23.5%	23.7%	23.6%	23.7%	23.5%
Highest fifth	42.6%	41.6%	42.0%	41.3%	41.6%
Top 5 percent	17.0%	16.8%	16.8%	15.8%	14.4%
Gini Index	.375%	.366%	.369%	.360%	.353%

SOURCE: Urban Systems Research & Engineering, Review
of Studies on the Measurement of Poverty,
prepared for the Department of Health, Education
and Welfare, September 1975.

The increased equality of family income is generally attributed to the increased incidence of two worker families,
and to the increased equality in the distribution of wealth
(assets) among families. A similar trend toward equality, however, is not evident from the data on the income of individuals. Indeed, individual incomes have become less equally distributed since World War II according to a number of studies.[*]
Both trends, toward equality in family incomes and inequality in individual incomes, have been small.

At the regional level, there have been much more significant changes in the distribution of income. During the 1960's,
Southern poverty has decreased due to industrialization, while
during the first half of the seventies farm incomes rose sub-

---

*See:
Edward Budd, "Postwar Changes in the Size Distribution of
Income in the United States," American Economic Review 60
(1970): 247-260.

Barry R. Chiswick & Jacob Mincer, "Time Series Changes in
Personal Income Inequality in the United States from 1934, with
Projections to 1985," Journal of Political Economy 80 (Supple-
ment; May/June 1972): 34-66.

Peter Henle, "Exploring the Distribution of Earned Income," Monthly Labor Review 95 (December 1972): 16-27

stantially. Since the distribution of income is crucially dependent on public policy, which is impossible to predict, we
have decided not to forecast changes in distributional patterns.

In sum, this study assumes that the relative proportions
of families in the ten income groups remains constant at both
the national and regional level.

Data on Other Socioeconomic Characteristics

In this study, we were also interested in the variation of
costs of the Act by the age and race of individuals. This distribution was determined by examining the correlation of race
and age with income, region of the country, and city size (the
three direct determinants of pollution costs). Distributional
data used in this analysis follows in Exhibit II-11 through
II-14.

Exhibit II-11 describes the income distribution in different age groups. According to this table, for families with
heads under age 25 years, the third income bracket is the
median, while for families with heads between ages 25 and 44,
it is the fifth income bracket. The earnings peak is attained
between ages 46 and 54. In general, the extremities of the
age distribution—the young and the old—earn the lowest levels
of income. If the population in a region has a relatively
high proportion of young and/or old people, the distribution
of income may be expected to be skewed towards the lower brackets. This is the case in the South. Exhibit II-12 shows the
age distribution of the population by region and by race. One
interesting feature of this table is that it shows a relatively low percentage of young Negro families and a relatively high
percentage of old Negro families living in the South. This may
be explained in terms of recent migration patterns: young Ne-
gros migrating from the rural South to the metropolitan North.

Differences in the distribution of income by race and city
size are described in Exhibit II-13 and by race and region in
Exhibit II-14. For Whites, the fifth income group represents
the median. For Negroes, the median income is much lower, falling in the third income group.

Comparison Data

In order to make a normative judgement about the distribution of the burden of water pollution abatement costs, some
yardstick was needed. In this study, we used two different
comparisons to judge the "goodness" of the burden distribution:
the Gini index and the tax structure.

23

Exhibit II-11

Distribution of Income by Age of Family Head

Percent of Families in Different Income Groups

Income Group	Age of Head of Family					
	under 25	25-34	35-44	45-54	55-64	over 65
1	25.6%	11.3%	8.9%	8.4%	12.4%	33.8%
2	16.7%	8.0%	6.3%	5.8%	8.6%	21.9%
3	19.8%*	13.7%	10.3%	9.5%	12.1%	13.6%*
4	14.1%	16.6%	13.4%	12.0%	13.3%	8.4%
5	13.9%	18.5%*	15.0%*	13.4%	14.5%*	8.0%
6	6.8%	20.3%	26.0%	25.7%*	20.2%	7.4%
7-10	3.1%	11.6%	20.1%	25.2%	18.9%	6.9%
TOTAL	100.0%	100.0%	100.0%	100.0%	100.0%	100.0%

SOURCE: Current Population Report, Series P-60, Number 97, Table 20, page 40.

*median income

The Gini index was first calculated assuming no water pollution abatement costs; this baseline Gini index (.27) was shown in the previous section. The Gini was then recalculated given the new distribution resulting from the Act. A decrease in the Gini is an indication of a reduction of inequality of income.

The distribution of the costs of the Act was also compared with the distribution of various tax burdens. In particular the federal tax, the state and local tax, the total tax, the personal income tax, and the property tax burdens were considered. The distribution of the federal, state and local, and total tax burdens are presented in Exhibit II-15 and II-16. Exhibit II-17 is a Lorenz curve of the tax distribution, constructed in the same way as the Lorenz curve for income distribution. The Federal personal income tax and the property tax burdens are presented in Exhibit II-18 and II-19. It was assumed that the federal tax structure represents society's view of the optimal distribution, and it would thus serve as a reasonable yardstick.

Exhibit II-12

Population Distribution Within Regions By Age
of Family Head and Race

Family Group (age of head)	Number of Families Residing in		
	U.S.	North & West	South
ALL FAMILIES			
Number (000)	54,359	36,838	17,521
% distribution			
14-24 years	7.7%	7.2%	8.8%
25-34 years	22.3%	22.2%	22.6%
35-44 years	19.4%	19.4%	19.3%
45-54 years	20.5%	21.1%	19.2%
55-64 years	15.7%	16.1%	14.9%
65+ years	14.4%	14.0%	15.2%
WHITE FAMILIES			
Number (000)	48,919	34,242	14,677
% distribution			
14-24 years	7.5%	7.0%	8.7%
25-34 years	22.0%	21.8%	22.5%
35-44 years	19.2%	19.1%	19.5%
45-54 years	20.7%	21.3%	19.5%
55-64 years	16.0%	16.5%	15.0%
65+ years	14.6%	14.4%	14.9%
NEGRO FAMILIES			
Number (000)	5,440	2,596	2,844
% distribution			
14-24 years	9.9%	10.6%	9.2%
25-34 years	25.3%	27.4%	23.3%
35-44 years	21.1%	23.8%	18.7%
45-54 years	17.9%	18.3%	17.6%
55-64 years	13.0%	11.1%	14.8%
65+ years	12.8%	8.7%	16.6%
MEDIAN AGE OF FAMILY HEAD	42.0	40.0	44.4

SOURCE: Current Population Survey, Series P-60,
Number 97, Table 73, p. 155.

25

Exhibit II-13

Distribution Within Racial Groups By Type of Residence and Income Group (1973)

Percent Distribution

INCOME GROUP	TOTAL U.S.	In Metropolitan Area Total	1,000,000 or more Total	1,000,000 or more in central cities	1,000,000 or more outside central cities	under 1,000,000 Total	under 1,000,000 in central cities	under 1,000,000 outside central cities	Outside Metropolitan Area Total	Outside Metropolitan Area Non-farm	Outside Metropolitan Area Farm
WHITE FAMILIES											
Number (000)	48,919	32,584	18,209	6,170	12,019	14,375	6,293	8,082	16,335	14,400	1,929
Percent in each income group	100%	100%	100%	100%	100%	100%	100%	100%	100%	100%	100%
1	12.5%	10.4%	9.8%	13.3%	7.9%	11.1%	13.5%	9.3%	16.6%	16.3%	19.9%
2	9.3%	8.2%	7.4%	10.0%	6.0%	9.6%	10.0%	9.1%	11.4%	11.1%	13.3%
3	12.1%	10.8%	10.1%	11.5%	9.2%	11.9%	12.8%	11.1%	15.0%	15.1%	13.9%
4	14.0%	13.4%	12.1%	13.0%	11.6%	15.0%	14.2%	15.7%	15.2%	15.2%	14.0%
5	14.3%	14.5%	14.0%	14.0%	14.4%	15.1%	14.4%	15.6%	14.1%	14.4%	10.7%
6	20.8%	23.0%	24.4%	20.7%	26.3%	21.1%	20.0%	21.8%	16.4%	16.7%	14.4%
7	8.7%	9.8%	10.5%	8.9%	11.6%	8.7%	8.2%	9.2%	6.6%	6.6%	6.4%
8	5.2%	6.2%	7.3%	5.7%	8.1%	4.9%	4.3%	5.4%	3.1%	2.9%	4.6%
9	2.5%	3.1%	3.6%	2.9%	4.0%	2.3%	2.2%	2.4%	2.2%	1.4%	2.3%
10	0.6%	0.6%	0.8%	0.7%	0.9%	0.4%	0.4%	0.4%	0.4%	0.3%	0.5%
MEDIAN INCOME	$12,595	$13,566	$14,333	$12,645	$15,192	$12,662	$12,111	$13,055	$10,788	$10,868	$10,162
NEGRO FAMILIES											
Number (000)	5,440	4,154	2,646	2,106	539	1,508	1,116	392	1,286	1,167	119
Percent in each income group	100%	100%	100%	100%	100%	100%	100%	100%	100%	100%	100%
1	34.1%	31.4%	29.2%	30.1%	25.1%	35.3%	36.9%	30.6%	42.8%	41.3%	56.6%
2	14.8%	14.3%	13.7%	14.3%	11.2%	15.1%	15.1%	15.1%	16.9%	17.3%	13.2%
3	14.4%	14.1%	13.4%	14.1%	11.2%	15.2%	15.1%	15.2%	16.0%	16.5%	10.9%
4	11.6%	11.8%	12.4%	12.3%	12.7%	11.3%	11.6%	10.4%	10.7%	11.5%	4.5%
5	9.3%	9.9%	10.0%	9.8%	11.0%	9.8%	9.0%	11.9%	7.3%	7.2%	8.0%
6	9.8%	11.3%	12.6%	11.6%	16.6%	8.8%	7.9%	11.9%	4.7%	4.6%	5.1%
7	3.0%	4.4%	4.9%	4.6%	6.8%	3.3%	3.0%	4.2%	1.5%	1.5%	1.7%
8	1.5%	2.0%	2.6%	2.2%	3.9%	0.9%	1.0%	0.5%	—	—	—
9	0.7%	0.8%	1.1%	0.9%	1.5%	0.3%	0.4%	0.2%	0.1%	0.1%	—
10	0.1%	0.1%	0.1%	0.1%	—	—	—	—	0.1%	0.1%	—
MEDIAN INCOME	$7,269	$7,779	$8,282	$7,912	$10,071	$7,053	$6,844	$8,007	$5,780	$5,905	$4,537

*Includes races other than White and Negro, not shown separately.

SOURCE: Current Population Report, Series P-60, Number 97, Table 24, pp. 46-47.

Exhibit II-14

Distribution of Negro Families in Each Region
By Income Group (1973)

Income Group	Total U.S.	REGION			
		Northeast	Central	South	West
NEGRO FAMILIES					
Number (000)	5,400	1,019	1,073	2,844	100
Percent in each income group	100.0%	100.0%	100.0%	100.0%	100.0%
1	34.1%	29.8%	28.2%	38.4%	29.8%
2	14.8%	15.3%	10.7%	16.7%	12.6%
3	14.4%	15.4%	13.1%	14.9%	13.6%
4	11.6%	10.3%	14.9%	11.0%	11.8%
5	9.3%	10.0%	10.7%	8.4%	10.1%
6	9.8%	11.9%	12.8%	7.0%	14.0%
7	3.0%	4.7%	5.3%	2.5%	5.3%
8	1.5%	1.9%	3.0%	0.7%	2.0%
9	0.7%	0.7%	1.2%	0.5%	0.8%
10	0.1%	---	0.1%	0.1%	---

SOURCE: Calculated from: Current Population Survey,
Series P-60, Number 97, Table 35, pp. 75-76

Exhibit II-15

Existing Tax Structure

Percent of Various Taxes Paid by Each Income Group

Income Group	Federal Taxes	State & Local Taxes	Total Taxes	Federal Personal Income Tax	Property Tax
1	2.7%	5.2%	3.5%	0.8%	6.9%
2	3.4%	5.3%	4.0%	1.2%	6.3%
3	6.0%	7.5%	6.5%	3.9%	7.9%
4	9.5%	11.3%	10.0%	7.2%	11.1%
5	12.1%	13.7%	12.6%	10.1%	13.1%
6	26.6%	26.7%	26.5%	26.3%	24.7%
7	13.6%	12.9%	13.5%	14.7%	11.7%
8	8.4%	7.1%	8.0%	10.3%	6.2%
9	7.5%	4.8%	6.7%	10.5%	5.2%
10	10.2%	5.4%	8.7%	15.0%	6.8%
Total	100.0%	100.0%	100.0%	100.0%	100.0%

SOURCE: Musgrave, Case & Leonard, op.cit.

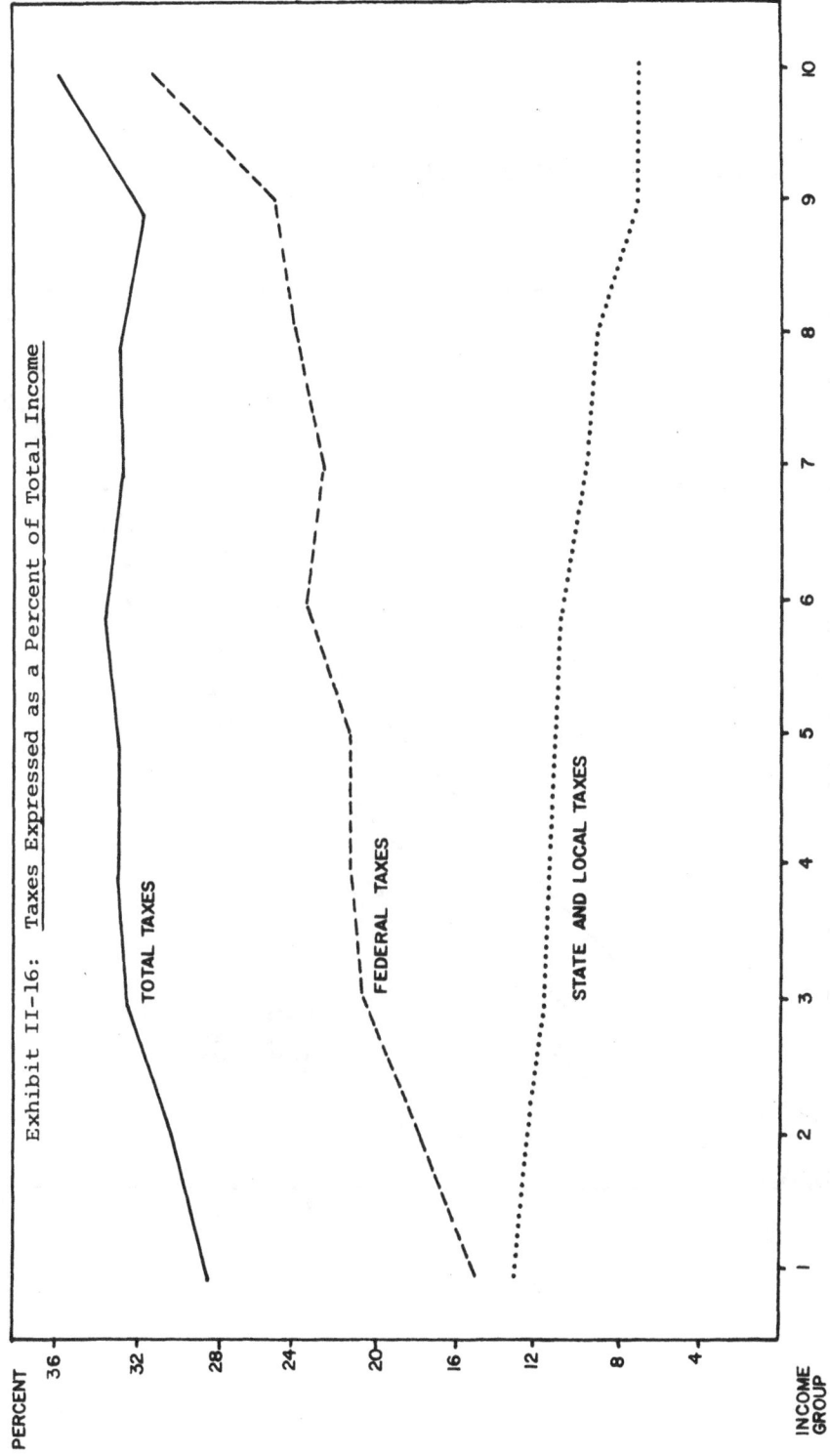

Exhibit II-16: Taxes Expressed as a Percent of Total Income

PERCENT

TOTAL TAXES

FEDERAL TAXES

STATE AND LOCAL TAXES

INCOME GROUP

29

Exhibit II-17

Lorenz Curves: The Distribution of Federal, State and Local Taxes
and the Total Tax Burden

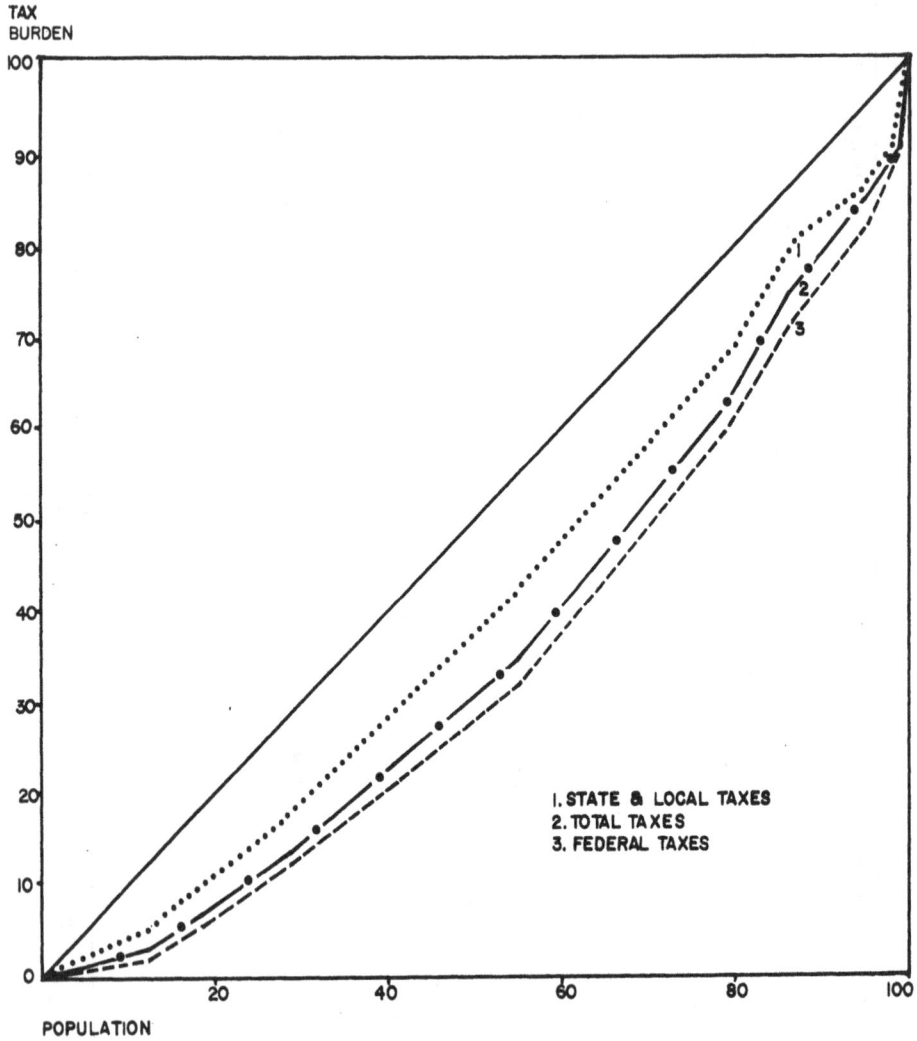

TAX
BURDEN

POPULATION

1. STATE & LOCAL TAXES
2. TOTAL TAXES
3. FEDERAL TAXES

30

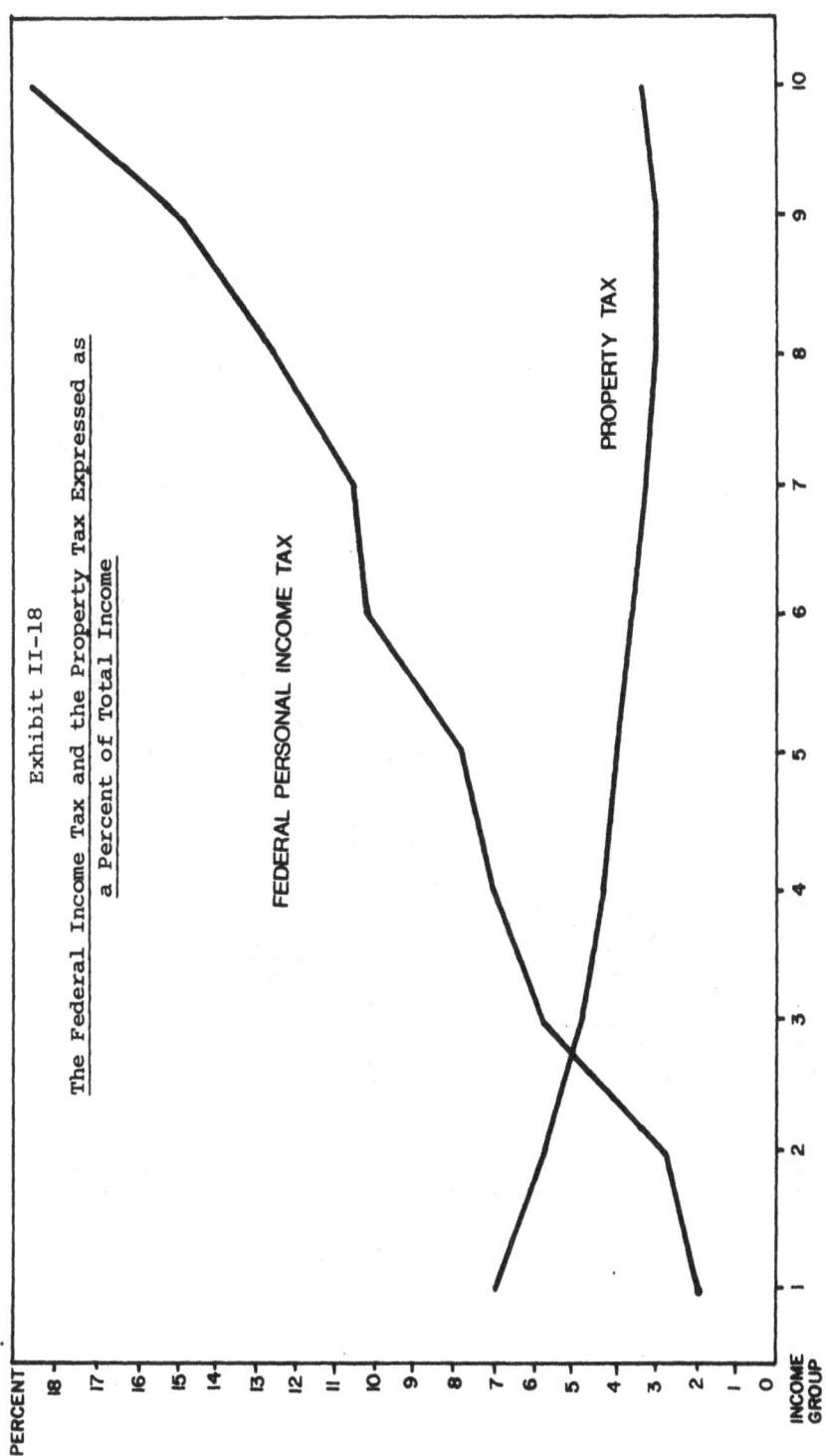

Exhibit II-18

The Federal Income Tax and the Property Tax Expressed as
a Percent of Total Income

FEDERAL PERSONAL INCOME TAX

PROPERTY TAX

31

Exhibit II-19

Lorenz Curve: The Distribution of the Property and
the Personal Income Tax

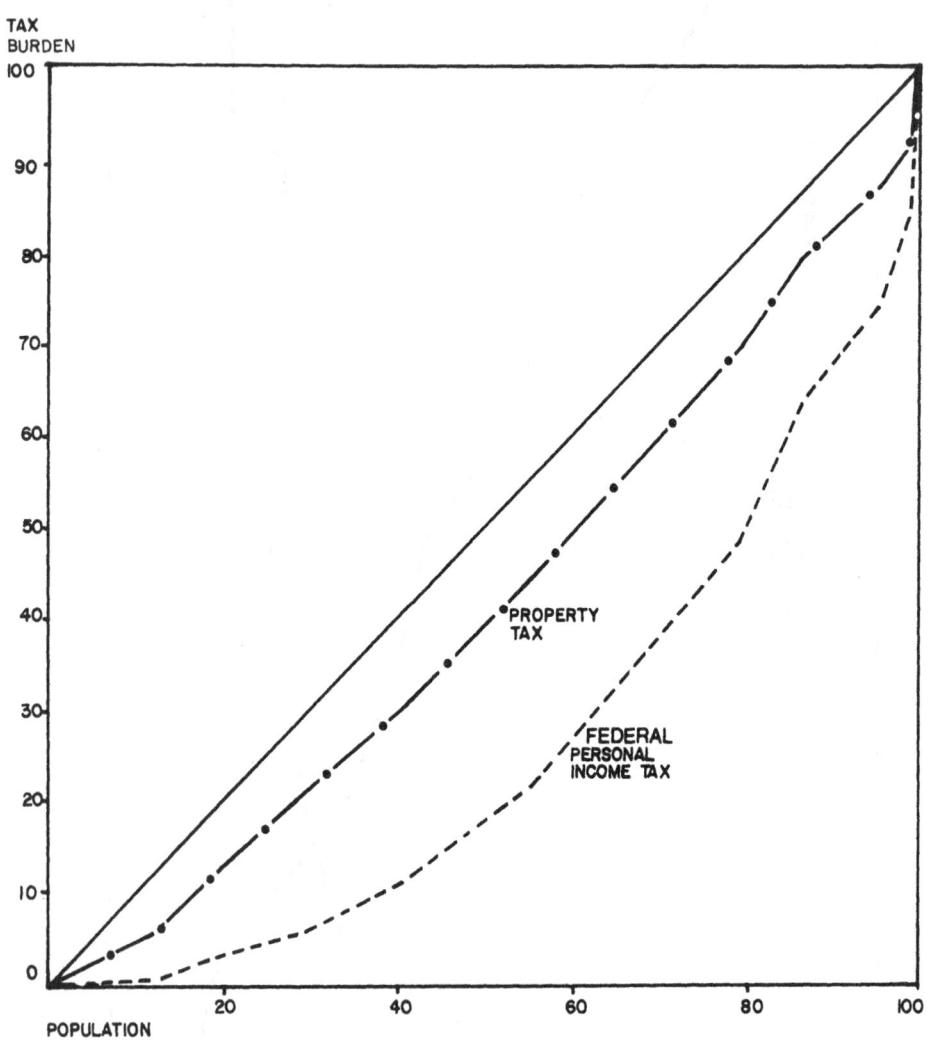

III. Municipal Expenditures

The Federal Water Pollution Control Act Amendments of 1972, P.L. 92-500, requires increasingly stringent levels of treatment on the part of all point source dischargers, including municipalities. Treatment costs will be met from government revenues, which in the main consist of taxes. As explained in Chapter I, the objective of this study is to examine the distribution of the fiscal burdens imposed by P.L. 92-500. The incidence of the necessary municipal expenditures depends on:

● their costs

● the methods of financing

● the incidence of the various taxes

● the distribution of income

Each of these factors is introduced below and analyzed in detail in the later sections of this Chapter.

The Costs: The treatment requirements of the Act, the state of technology, population size, resource costs, and the industrial use of municipal facilities are some of the factors which influence municipal treatment costs. The magnitude of these costs, in turn, determine the size of the burden which must be shared by the different groups in society.

The Methods of Financing: The treatment costs necessitated by the Act are shared among the three levels of government: federal, state, and local. According to Section 201 of the Act, the federal government contributes 75% of the capital costs of the treatment/interceptor facilities, as approved by EPA. It makes no contributions towards operation and maintenance (O&M) costs. Many states also provide fiscal assistance for capital costs; however, only in New York is a proportion (one-third) of the O&M costs financed by the state. The local share of the costs must be raised by the municipal, county or special district governments. The distribution of costs between the three levels of government will clearly influence the ultimate incidence pattern.

In addition, each level of government can raise the necessary revenues from a number of alternative sources, or taxes, such as personal and corporate income taxes, sales and excise taxes, property taxes, and so on. The choice of taxes will also influence the resulting incidence pattern. The federal government collects approximately 65% of its revenues from the individual and corporation income taxes, state governments collect over half their revenues from sales and excise taxes, while property taxes comprise 82% of revenue in municipal budgets.

To raise the necessary funds, governments may also cut back on other expenditures. For example, the funds earmarked for a new recreation facility may be used towards the construction of a municipal treatment plant. The group of people who would have used the recreation facility are then worse off, and are, in fact, paying a part of the treatment plant costs.

It should be pointed out that "benefits" as used in this methodology is assumed to mean expenditures on behalf of particular income groups. This is a standard public finance technique; it eliminates the thorny problem of evaluating the real benefits of expenditures such as health care, education, or recreation.

The Incidence of Various Taxes/Expenditures: The incidence of particular taxes and expenditures depends upon the initial recipients of the taxes and expenditures as well as upon the response or adjustment they cause in the economy. The corporate income tax presents one example of the adjustment process. The response of the corporations in many cases may be to try to raise their prices in order to maintain their profits. To the extent that they succeed in doing this, they "shift" part of the tax burden to the consumers. The assumptions concerning the adjustment process are called "shifting assumptions." According to one set of shifting assumptions,* one-third of the corporate income is passed along to consumers

---

*Roger A. Herriot and Herman P. Miller, "Tax Changes Among Income Groups: 1962-1968," Business Horizons, February 1972.

34

in the form of higher prices, and the remaining two-thirds is borne by the owners of capital. Alternative assumptions about the adjustment and shifting process are possible; for example, the entire tax burden is borne by the owners of capital.* The incidence of the tax then depends on the adjustment or shifting process, as well as on the distribution of consumption and capital ownership among the different income groups.

The incidence of various categories of expenditures depends on initial "recipients" of expenditures (or users of the goods and services provided). For example, the benefits of expenditures on education are generally imputed to the attending students. The benefits are assumed to stay with the initial recipients; in other words, no shifting or adjustment takes place in any expenditure category.

The Distribution of Income: Finally, the incidence of pollution expenditures depends upon the distribution of income. To the extent that the federal government finances its expenditures from the individual income tax, any individual taxpayer's share, or burden, depends upon his income. Similarly, the incidence of the various tax measures and expenditures has been estimated for a number of income groups. Any individual's share in the tax burden, or benefit loss, depends upon his income. Thus, the incidence of pollution control expenditures depends upon the distribution of income.

The incidence of treatment cost burdens is depicted in Exhibit III-1. The costs (capital or O&M) are divided among the three levels of government: federal, state, and local. Each level of government has a number of options for financing its expenditures: through the imposition of a variety of taxes or through expenditure cuts. The incidence of each varies; therefore, the ultimate incidence pattern depends upon the "mix" of taxes and expenditure cuts selected.

The incidence of one financing method is depicted in the following diagram--that of the general sales tax. The first income distribution line describes the cumulative percentage of the population falling below the upper limit of each of the 10 income groups. The second income distribution line represents a similar presentation for the cumulative distribution of total income. The lowest income group pays 5.7% of all the general sales taxes collected. Although this burden falls

_____

*Richard A. Musgrave, Karl E. Case, and Herman Leonard, "The Distribution of Fiscal Burdens and Benefits," Public Finance Quarterly, July 1974.

Exhibit III-1:: The Incidence of Municipal Treatment Costs

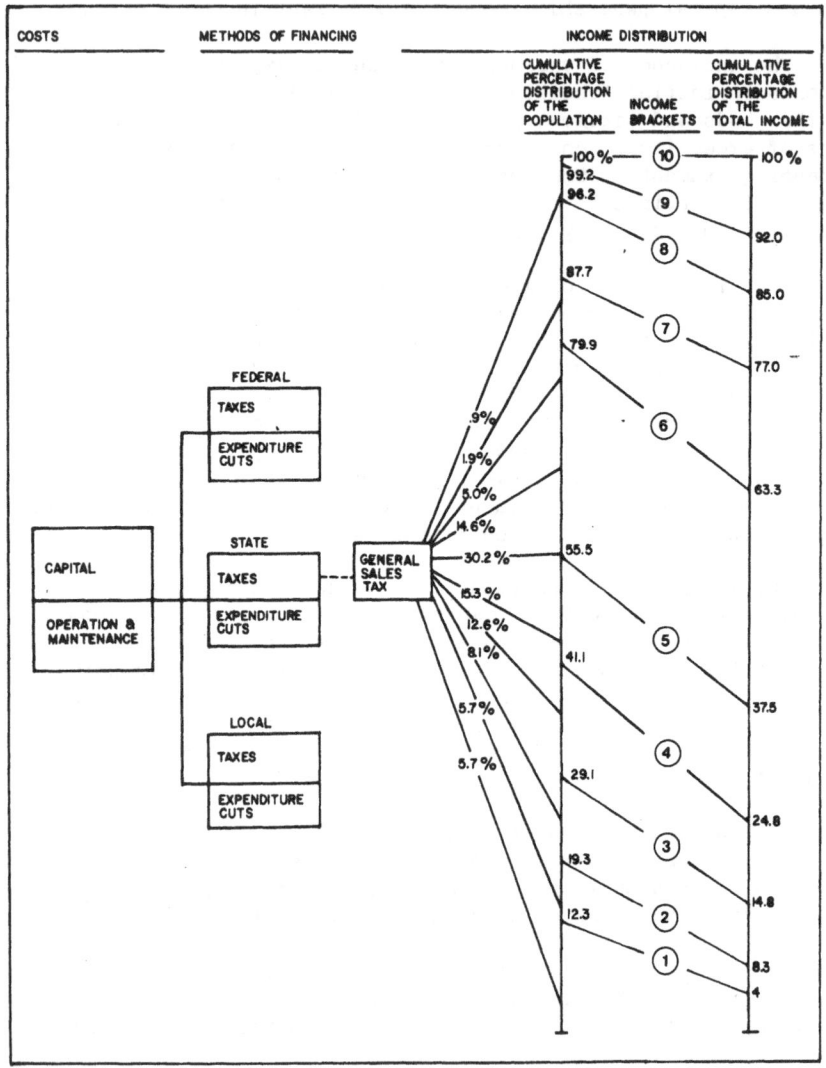

on 12.3% of the population, they receive only 4% of total
income. In the second income group, another 5.7% of the tax
is paid, with the burden falling on 7.0% of the population
with 4.3% of the total income. The general sales tax, then,
represents a disproportionately heavy burden for the lower
income groups and a disproportionately light burden for the
higher brackets. It is, therefore, said to be regressive.
Similar examples could be worked out for each financing alternative, such as the federal corporation income tax, the local
property tax, and so on.

Since the incidence pattern depends upon the assumptions
concerning the magnitude of the costs, the methods of financing selected, the impact of the methods on different income
groups, and on the distribution of income, a large number of
incidence patterns may be calculated as these assumptions are
varied. The present analysis uses a small number of alternative scenarios and calculates the resulting incidence pattern
for each. These scenarios are based on alternative assumptions concerning:

- the magnitude of water pollution control
  costs;

- the methods through which they will be
  financed;

- the incidence of various taxes/expenditure
  cuts.

The results are presented later in this chapter.

THE COSTS OF MUNICIPAL TREATMENT FACILITIES

The first step required to examine the distribution of
the cost burdens imposed by P.L. 92-500 is to estimate the
magnitude of the costs. This section will review the cost
estimates used in the incidence analysis. In addition, it
will discuss the concept of baseline expenditures, that is,
the water pollution control expenditures which would have
taken place in the absence of the Act.

The Costs of Municipal Facilities under P.L. 92-500

In sections 516(b) and 205(a), P.L. 92-500 required the
conduct of a joint state-EPA survey of estimates of construction costs for publicly owned waste treatment works. The
resulting 1974 Needs Survey represents EPA's official estimates of the costs of meeting the requirements of P.L. 92-500.
A summary of the Needs Survey estimates is presented in

37

Exhibit III-2. In this table, the categories are as follows:*

Category I: Costs for facilities which would provide "best practicable wastewater treatment technology (BPWTT).

Category II: Costs for treatment facilities that must achieve more stringent levels of treatment.

Category IIIA: Costs for the correction of infiltration/inflow problems.

Category IIIB: Costs for replacement and/or major rehabilitation of existing sewage collection facilities.

Category IVA: Costs for construction of collector sewer systems designed to correct violations caused by rain discharges, seepage to waters from septic tanks, and so on.

Category IVB: Costs of new interceptor sewers and transmission pumping stations necessary for the bulk transport of wastewaters.

Category V: Costs for preventing the periodic bypassing of untreated wastes from combined sewers to an extent violating water quality standards and effluent limitations.

The costs of Categories I, II, and IVB total $46 billion. These elements represent the traditional water quality program of treatment plants and interceptors. Categories III, IVA, and V total another $61 billion. In addition, the survey considered Category VI, the costs associated with the treatment and/or control of stormwaters. This category alone amounted to some $35 billion. The distribution of the cost burdens was calculated for Categories I, II, and IVB, and for Categories I-V.

Although the Needs Survey costs are inaccurate estimates** of the costs of meeting the requirements of P.L. 92-500, this report uses the Needs Survey figures in estimating the

---

*U.S. Environmental Protection Agency, Cost Estimates for Construction of Publicly Owned Wastewater Treatment Facilities, 1974 Needs Survey, February 1975.

**APWA, A Study of the Costs Associated with the Meeting of the Requirements of P.L. 92-500, December 1974, prepared for the NCWQ; and Metcalf & Eddy, Inc., Assessment of Technologies and Costs for POTW's under P.L. 92-500, April 1975, prepared for the NCWQ.

distribution of costs. This was done for a number of reasons. First, the survey results represent EPA's official estimates of the costs of meeting the requirements of the Act. Second, and more important, grants are currently allocated on the basis of the Needs Survey. Therefore, although the magnitude of the expenditures undertaken may be substantially below the Needs Survey estimates, their distribution between states will be the same as the amounts reported here--unless the allocation formula is changed. For the purposes of the analysis performed under this contract, the state by state distribution of costs was far more important than the actual amount of the costs. This is the case because the various state and local governments use different fiscal instruments to finance their share. Of pollution control expenditures, state by state differences in fiscal structures will be discussed in more detail later in this chapter. Alternative cost estimates may be used to scale the results obtained here. In other words, in the scenario where water pollution control costs amount to a half of the Needs Survey amounts for, say, Categories I-V, the per family burden for each group should be halved, but the distribution of the burden over the income groups or geographical areas would remain unchanged.

Alternative cost estimates have been estimated by Metcalf and Eddy. According to their calculations, the total cost of meeting the Needs reported in Categories I and II of the Needs Survey is $26.3 billion, with $20.3 billion required for secondary treatment and the remaining $6.0 billion for the more stringent levels. These figures are comparable to $12.6 billion for secondary treatment and $15.7 billion for more stringent treatment according to the Needs Survey.

The Commission also estimated the costs imposed by the Act in a variety of scenarios--which include alternative allocation formulas, timing schedules for expenditures, and alternative levels of federal funding for the eligible categories.

The operations and maintenance (O&M) costs which must be incurred in addition to the capital expenditures estimated by the Needs Survey are particularly relevant for distributional considerations. Typically, O&M costs are financed from different sources than capital expenditures. There is no federal assistance available for O&M charges, and only New York has a state O&M assistance program. O&M charges are almost exclusively financed from local sources, with an increasing reliance on user charges. As the user charge is a particularly regressive fiscal instrument, the magnitude of the O&M charges has significant distributional consequences.

Exhibit III-2

Costs Reported for Construction of Publicly-Owned Wastewater Treatment Facilities (Millions of 1973 Dollars)

REGION State	Total	CATEGORY						
		I	II	IIIA	IIIB	IVA	IVB	V
NORTHEAST								
New England								
Maine	$ 575	$ 130	$ 4	$ 1	$ 2	$ 102	$ 139	$ 96
New Hampshire	740	125	95	23	0	184	164	147
Vermont	204	31	60	9	0	33	34	34
Massachusetts	2,964	587	205	33	35	759	533	824
Connecticut	1,594	171	94	21	38	311	213	738
Rhode Island	447	45	34	3	0	169	108	87
Middle Atlantic								
New York	15,302	1,663	994	509	2,341	2,845	1,946	5,004
New Jersey	4,894	1,127	571	199	174	540	904	1,377
Pennsylvania	5,454	742	265	91	48	1,246	622	2,439
Delaware	546	65	34	100	67	65	100	114
NORTH CENTRAL								
East North Central								
Ohio	7,773	214	1,302	637	114	592	851	4,062
Indiana	2,903	200	348	151	185	332	289	1,397
Illinois	6,234	314	1,635	176	60	466	394	3,188
Wisconsin	2,044	164	373	79	134	382	402	508
Michigan	8,102	106	598	96	456	992	969	4,884
West North Central								
Minnesota	1,330	59	415	52	14	188	233	368
Iowa	911	132	178	123	9	121	222	125
Missouri	2,298	400	63	219	0	358	380	878
North Dakota	189	51	0	1	0	63	23	50
South Dakota	75	33	32	1	0	2	7	0
Nebraska	924	144	18	38	1	26	65	630
Kansas	1,783	185	67	536	294	316	272	113
SOUTH								
South Atlantic								
Maryland	3,642	10	1,465	51	1,132	107	849	27
West Virginia	2,360	91	151	145	5	695	1,078	194
Virginia	1,884	364	317	290	32	226	448	206

Exhibit III-2 (Cont'd)

Costs Reported for Construction of Publicly-Owned Wastewater Treatment Facilities (Millions of 1973 Dollars)

REGION State	Total	I	II	IIIA	IIIB	IVA	IVB	V
				CATEGORY				
North Carolina	$1,480	211	$451	$52	$0	$383	$382	$0
South Carolina	977	333	45	21	0	227	350	0
Georgia	1,519	200	501	12	2	200	319	283
Florida	2,704	566	452	39	44	746	856	0
East South Central								
Kentucky	1,824	51	243	53	63	406	355	652
Tennessee	1,210	112	293	45	39	332	272	114
Mississippi	494	35	202	45	31	57	122	1
Alabama	778	114	165	102	11	192	193	0
West South Central								
Arkansas	898	0	331	68	0	246	251	1
Louisiana	1,283	302	6	352	97	333	191	0
Oklahoma	1,484	0	406	55	551	214	256	0
Texas	3,222	9	1,617	207	214	657	399	118
WEST								
Mountain								
Montana	127	39	16	5	1	21	35	7
Wyoming	84	40	0	0	0	28	15	0
Colorado	409	78	66	29	24	74	115	23
New Mexico	155	52	7	3	1	54	38	0
Idaho	393	46	71	22	16	102	99	35
Nevada	209	30	102	1	0	31	45	0
Utah	291	195	0	14	2	57	23	0
Arizona	500	159	8	1	1	231	99	0
Pacific								
Washington	1,836	283	56	91	454	299	336	316
Oregon	1,081	146	1	56	324	130	161	262
California	6,208	1,713	1,242	343	47	812	1,149	902
Alaska	405	234	0	4	1	70	85	9
Hawaii	523	203	27	0	0	84	209	0
D.C.	1,052	0	67	40	216	0	2	726
TOTAL	$107,314	$12,635	$15,720	$5,287	$7,287	$17,458	$17,849	$31,076

SOURCE: U.S. Environmental Protection Agency, Cost Estimates for Construction of Publicly Owned Wastewater Treatment Facilities, 1974 Needs Survey, February 1975.

41

The relationship O&M charges to capital costs depends on the type of capital expenditure undertaken: the O&M costs of interceptor and collector sewers are substantially lower than the O&M costs of treatment plants. Further, even for treatment plants there is considerable variation in this ratio depending upon the type of treatment technology. In this analysis we have used the estimates derived by Metcalf and Eddy.* These are: for treatment plant expenditures, annual O&M cost of 3.25% of capital; and for sewer operating expenditures 0.84% of capital. For categories I, II and IVB, this implied 2.43%, and for categories I-V, 1.48%. These estimates are below those obtained in the survey of local governments financing water pollution abatement expenditures. For the 226 projects reviewed in the survey, the average ratio of O&M charges to capital costs was 3.96%, with the mix of projects including both treatment plants and sewer projects.

Baseline Costs

Baseline expenditures may be defined as the water pollution control expenditures which would have occurred in the absence of P.L. 92-500. Ignoring baselines, the distribution of water pollution control expenditures under the Act are estimated. If baseline expenditures are included in the analyses, the differential incidence is estimated, that is, the distribution of the burden resulting from the additional costs and changes in financing imposed by the Act. This consideration is particularly relevant in view of the fact that the Act not only imposed additional costs, but also provided for a substantially increased level of federal assistance.

Between 1957 and 1972, less than 15% of the total costs of the construction of municipal treatment facilities was financed from federal sources.** According to our survey, for the projects which did receive federal assistance, this proportion was substantially higher. Between 1967 and 1972, projects with total costs in excess of $1,000,000 received 44% federal assistance under P.L. 84-660. Similar sized projects under P.L. 92-500 received approximately 75% federal assistance. Federal participation in the total program rose from 15% to approximately 31% for 1973 to 1974.**

---

*Metcalf & Eddy, Inc., Draft, "Report to National Commission on Water Quality on Assessments of Technologies and Costs for Publicly Owned Treatment Works under Public Law 92-500," April 1975.

**U.S. Environmental Protection Agency, Review of the Municipal Waste Water Treatment Works Program, November 1974, Washington, D.C.

Since federal funding is generally financed from more progressive fiscal instruments than local funds, a shift in the relative proportions of the cost burden from local to federal sources should result in a more progressive distribution of the (larger) burden.

Baseline expenditures were projected by the Commission on the basis of past trends. Expenditures* on sewer systems began in the past century, with the first sanitary sewer begun in Chicago in 1855. By the 1930's, approximately half the population lived in sewered areas and 34% of the sewered population was receiving some wastewater treatment. By 1962, these percentages were up to 73% and 86% respectively. During the decade of the 1960's, this growth has leveled off somewhat.

On the basis of these trends, the Commission developed two baselines.** The high baseline was derived from an extrapolation of past trends in contract awards for treatment plant, interceptor and collection sewer construction. The low baseline assumes that the present trend in sewer expenditures--representing the bulk of the baseline expenditures--would level off once the majority of the urban population was sewered. The high and the low baseline expenditures on water pollution control are presented in Exhibit III-3.

FEDERAL, STATE AND LOCAL GOVERNMENT FISCAL STRUCTURES

The purpose of this section is to describe the main fiscal instruments used by the federal, state and local governments and to summarize their evolution over the recent past. This will set the stage for the forecasts of how impacted government units will finance the costs of the public wastewater treatment facilities necessitated by P.L. 92-500. These forecasts will be described in more detail later in this chapter. Another section of this chapter will examine the distributional impacts of the fiscal instruments described in this section. The analysis presented here will make a distinction between different types of governmental units involved in the provision of public wastewater facilities, since they have differing responsibilities and financial structures. The government units involved are the federal government, the state governments, county governments, municipalities and townships, and special sewer districts. The last three

---

*U.S. Environmental Protection Agency, The Economics of Clean Water, 1973, Chapter III.

**National Commission on Water Quality, Draft Report, Chapter VI, September 1975.

43

EXHIBIT III-3

Estimates of Expenditures on Sewerage
Without P.L. 92-500
(Millions of 1973 $s)

Year	State and Local Expenditures		Total Expenditures Including Federal*	
	High Baselines	Low Baselines	High Baselines	Low Baselines
1974	$2.01	$2.22	$2.36	$2.61
1975	2.06	2.02	2.42	2.38
1976	2.11	1.77	2.48	2.08
1977	2.19	1.40	2.58	1.65
1978	2.24	1.40	2.64	1.65
1979	2.31	1.40	2.72	1.65
1980	2.36	1.40	2.78	1.65
1981	2.42	1.40	2.85	1.65
1982	2.48	1.40	2.92	1.65
1983	2.54	1.40	2.99	1.65
1984	2.54	1.40	2.99	1.65
1985	2.54	1.40	2.99	1.65

*Assuming a 15% rate of federal participation.

SOURCE: National Water Quality Commission, Draft Final Report, September, 1975.

governmental units are involved in both operating and financing public wastewater treatment facilities. In contrast, the
federal government and the state governments do not have any
direct operating responsibility, but they have a role in the
financing of public sewer systems through intergovernmental
revenue transfers to the other units of government.

There are two ways to measure the relative importance of
county governments, municipalities, and special districts in
the provision of public sewerage services. One method is to
examine the distribution of direct expenditures for sewerage
among the governmental units. This can only be done for years
in which a Census of Government was held.* Exhibits III-4 and
III-5 show the distribution of direct expenditures for sewerage among local government units in 1962, 1967, and 1972 for
the U.S. as a whole.** It should be noted that there is no
specific data on special districts in the 1962 Census, nor is
there a breakdown of county government expenditures for sewerage between capital and non-capital outlays in that Census.
The most interesting item is the tremendous increase in the
county government sewerage expenditures between 1967 and
1972, which led to a doubling of their share of total local
government sewerage expenditures.

The alternative approach is to examine the distribution
of financial burdens rather than expenditures among the
various levels of government--in effect, we should modify the
data in Exhibits III-4 and III-5 to allow for the different
amounts of intergovernmental transfers for sewerage received
by county governments, municipalities, and special districts.
The necessary data is, unfortunately, not available. As a
proxy, however, the different amounts of sewer bonds issued
by the various units of government could be considered. This
is a highly imperfect substitute because it reflects variations in the use of long-term bonds as a financial instrument
as well as variations in capital expenditures and receipts of

---

*The EPA and its predecessor agencies have published data
on public sewerage construction activities; however, none of
the publications distinguishes between county governments,
municipalities, and special districts.

**All the Census data is for fiscal years. For conve-
nience the label "fiscal year" will be dropped. Thus all
references in this section to the Censuses of Governments in
1962, 1967, and 1972 are to be taken as denoting the fiscal
years 1961/62, 1966/67 and 1971/72 respectively.

EXHIBIT III-4

Total Direct Expenditures for Sewerage and Growth Trends by Type of Governmental Unit: 1962, 1967, and 1972

Governmental Unit	Amount ($Million)			% Distribution			% Increase
	1962	1967	1972	1962	1967	1972	1967 – 1972
County Government	$ 91.7	$ 112	$479.7	8.8%	6.7%	14.4%	+328.3%
Municipality	951.2	1218	2194.4	91.2	72.7	65.8	+ 80.2%
Special Sewer District	NA	345	661.5	NA	20.6	19.8	+ 91.7%
TOTAL:	$1042.9	$1675	$3335.6	100.0%	100.0%	100.0%	+ 99.1%

SOURCE: 1962 Census of Governments, Vol. 4, No. 2, Table 1; No. 3, Tables 11,19
1967 Census of Governments, Vol. 4, No. 2, Table 1; No. 3, Table 17; No. 4, Tables 11,19
1972 Census of Governments, Vol. 4, No. 2, Table 1; No. 3, Table 17; No. 4, Tables 11,19

EXHIBIT III-5

Capital Outlays for Sewerage,
by Type of Governmental Unit: 1967 and 1972

Governmental Unit	Amount ($Million)		% Distribution		% Increase
	1967	1972	1967	1972	1967-1972
Municipality	$ 760.0	$1377.1	71.1%	64.3%	+ 81.2%
Special Sewer District	234.6	445.3	22.0%	20.8%	+ 89.8%
County	74.0	320.4	6.9%	15.0%	+333.0%
Total:	$1068.6	$2142.8	100.0%	100.0%	+100.5%

SOURCE: Same as for Exhibit III-4.

intergovernmental transfers.* Moreover, we only have separate
data on sewer bond sales for the years up to 1966; thereafter,
the available data pertains to water and sewer bonds jointly.
Exhibits III-6 and III-7 show the distribution of bond sales
among different issuing authorities in 1962, 1966, and 1972.
Two features of the results stand out. One is the rising
volume of state bond activity both by the state governments
themselves and through statutory authorities, which is a
reflection of the growing state government involvement in
financing public water and sewer services. The other is the
increase in the county government's bond activity in the late
1960's, which is consistent with data on sewerage expenditures
in Exhibits III-4 and III-5.

Because of the differences in data sources, Exhibits III-5
and III-7 are not directly comparable. Nevertheless, they
present an intriguing contrast: municipalities accounted for

---

*The original data for the period up to 1966 distinguishes
between water bonds, sewer bonds and combination water and
sewer bonds. The FWPCA report, The Cost of Clean Water:
Economic Impact on Affected Units of Government (January 1968),
suggests that 70% of the combination bonds be attributed to
sewerage. The reason for this suggestion is never made clear.
In the absence of any better information, we have adopted it
here in calculating the total volume of sewerage bond sales.

Exhibit III-6

Water and Sewer Bond Sales,
by Issuing Authority: 1962, 1966, and 1972

Issuing Authority	SEWER BONDS Amount ($ Million)		WATER & SEWER BONDS Amount ($ Million)		% Increase
	1962	1966	1966	1972	1966-72
State	$ 0	$ 13.5	$ 222.7	$ 443	+ 98.9%
County	43.5	38.2	50.6	339	+570.0%
Municipality & Township	389.4	365.3	566.4	869	+ 53.4%
Special District	151.9	139.6	306.5	498	+ 62.5%
Statutory Authority	84.2	88.2	285.6	292	+ 2.2%
Total	$669.2	$644.9	$1431.8	$2441	+ 70.5%

SOURCE: Investment Bankers Association, Quarterly
Municipal Statistical Bulletin, various issues.

Exhibit III-7

Percentage Distribution of Water and Sewer Bond
Sales, by Local Government Issuing Authority:
1962, 1966, and 1972

Issuing Authority	SEWER BONDS		WATER & SEWER BONDS	
	1962	1966	1966	1972
County	7.4%	7.0%	5.5%	19.9%
Municipality & Township	66.6%	67.3%	61.3%	50.9%
Special District	26.0%	25.7%	33.2%	29.2%
Total	100.0%	100.0%	100.0%	100.0%

SOURCE: Same as Exhibit III-6.

a significantly larger share of sewerage expenditures in 1972 than of bond sales; the opposite is true of county governments and special districts. This phenomenon is consistent with either of two hypotheses: that intergovernmental revenues for sewerage go disproportionately more to municipalities, or that municipalities rely less on bond issues for financing their sewerage capital projects than the other local government units. According to our analysis of federal grant assistance for publicy owned wastewater treatment facilities, and the results of the survey conducted among the grant recipients (described later in this chapter), municipalities have received a disporportionate share of the grant assistance.

In this connection, it may be useful to note that there is an ambiguity in the term "financing" as it appears in most discussions of financing pollution abatement expenditures. One should make dinstinctions between the initial financing of a construction project and its ultimate financing. The former concept refers to the immediate sources of funding for the project--typically the proceeds of bond sales, current revenues (including intergovernmental transfers), or accumulated reserves. The latter concept refers to the ultimate sources of these funds--in the case of bonds, for example, or reserve funds, the term refers to the tax or other revenues from which debt service charges are paid or reserves accumulated. The balance of this section will examine the various revenue sources which might constitute the ultimate source of finances for public sewerage expenditures by the federal, state and local governments. The role of bond financing as an initial source of funds will also be examined, as well as the constraints on its use which are imposed by statutory debt limits and interest rate limitations.

Federal Government Revenue Sources

The distribution of federal government revenues among alternative revenue sources in 1960 and 1972 is shown in Exhibit III-8. The most important trend in this period has been the growth in revenues from payroll taxes earmarked for social security programs such as OASDI. The individual income tax remains the largest single revenue source, providing 45.4% of total federal revenues in 1972; the corporation net income tax accounted for an additional 15.4% of revenues in that year. The other major sources of revenue were excise taxes levied on alcohol, tobacco, gasoline and diesel fuel, oil, tires, trucks and parts, passenger and freight air transportation, interstate telephone calls, and other miscellaneous items. Together with customs duties, these accounted for 9% of federal revenues in 1972.

```
                        Exhibit III-8

              Federal Government Revenue Sources and
                 Growth Trends:   1960 and 1972


                          Amount         %          %
                        ($ Billions)  Distribution  Increase
Revenue Source           1960   1972   1960   1972  1960-72


Personal Income Tax     $40.7  $ 94.7  44.0%  45.4%  +132.7%
Corporation Net Income
  Tax                    21.5   32.2   23.2%  15.4%  + 49.8%
Social Security Pay-
  roll Taxes             14.7   53.9   15.9%  25.8%  +266.7%
Excise Taxes             11.6   15.5   12.5%   7.4%  + 33.6%
Estate & Gift Taxes       1.6    5.4    1.7%   2.6%  +237.5%
Customs Duties            1.1    3.3    1.2%   1.6%  +200.0%
Miscellaneous             1.2    3.6    1.3%   1.7%  +200.0%


TOTAL                   $92.5  $208.6  100.0% 100.0% +125.5%


SOURCE:   Budget of the United States Government
```

State Government Finances

The single largest source of revenues for state governments is represented by tax receipts, followed by intergovernmental revenues primarily from the federal government.*
Other sources include revenues from current charges and miscellaneous revenues (consisting of rents, royalties, fines,
interest earnings, special assessments, and sale of property).
These revenues taken together are referred to as "general
revenues." In addition, there are revenues from state liquor
stores and from insurance trust systems (employee retirement,
unemployment and workman's compensation); together with
general revenues, these comprise the total revenue of the
state governments. Exhibit III-9 shows distribution of total
revenues among these sources in 1962 and 1972. For the decade
as a whole, total revenue increased at an average annual rate
of 11.6%. The largest increase was registered by intergovernmental revenues, reflecting the massive growth in the federal
government's grant programs in the areas of housing and community development, welfare, and education. Revenues from
current charges and miscellaneous general revenues also grew
appreciably faster than total revenue as a whole, although

_____

*Transfers from local government account for only about 4%
of the intergovernmental revenues of state governments.

50

Exhibit III-9

State Government Revenue Sources and Growth Trends:
1962 and 1972

| Revenue Source | Amount ($ Millon) | | % Distribution | | % Increase |
	1962	1972	1962	1972	1962-72
Intergovernmental Revenues	$ 7,480	$ 27,981	19.9%	24.9%	+274.1%
Tax Revenues	20,561	59,870	54.7%	53.3%	+191.2%
Current Charges	2,198	7,820	5.8%	7.0%	+255.8%
Miscellaneous General Revenues	918	2,960	2.4%	2.6%	+222.4%
Liquor Store Revenues	1,134	1,904	3.0%	1.7%	+ 67.9%
Insurance Trust Fund Revenues	5,304	11,773	14.1%	10.5%	+122.0%
Total Revenue	$37,595	$112,309	100.0%	100.0%	+198.7%

NOTE: Because of rounding, details may not add up to totals.

SOURCE: 1962 Census of Governments, Vol. IV, No. 4, Compendium of Government Finances, Tables 1, 3.

U.S. Department of Commerce, Bureau of the Census, State Government Finances in 1972, 6F72 No. 3, Table 7.

these are still relatively minor components of state government revenues.

Exhibit III-10 shows the breakdown of state government tax receipts in 1962 and 1972 among the following categories: general sales tax, selective sales tax (levied primarily on motor fuel, which accounts for about 45% of selective sales tax receipts, and on alcoholic beverages, tobacco products, insurance and public utilities); license taxes (about 60% of the receipts come from motor vehicle and driving licenses); individual income taxes; corporation net income taxes; property taxes; death and gift taxes; severance taxes; and miscellaneous other taxes (including poll taxes and documentary and stock transfer taxes). The tables indicated that there were some significant changes in state tax structures during this period, although for the most part this represented a continuation of the trends which had been in evidence throughout the post-World War II period. During the 1960's receipts from

Exhibit III-10

State Government Tax Collections in Detail
and Growth Trends:
1962 and 1972

Tax Source	Amount ($ Million)		% Distribution		% Increase 1962-72
	1962	1972	1962	1972	
General Sales & Gross Receipts	$ 5,111	$17,619	24.9%	29.4%	+244.7%
Selective Sales & Gross Receipts	6,927	15,631	33.7%	26.1%	+125.7%
Licenses	2,175	5,374	10.6%	9.0%	+147.1%
Individual Income	2,728	12,996	13.3%	21.7%	+376.4%
Corporation Net Income	1,308	4,416	6.4%	7.4%	+237.6%
Property	640	1,257	3.1%	2.1%	+ 96.4%
Death and Gift	516	1,294	2.5%	2.2%	+150.8%
Severance	451	758	2.2%	1.3%	+ 68.1%
Other	705	525	3.4%	0.9%	- 25.5%
TOTAL	$20,561	$59,870	100.0%	100.0%	+191.2%

Note: Because of rounding, details may not add up to totals.

SOURCE: Same as for Exhibit III-9.

general sales taxes for the first time exceeded receipts from
selective sales taxes, making this the largest source of tax
revenues for state governments. At the present, only five
states are without a general sales tax--Alaska, Delaware,
Montana, New Hampshire, and Oregon. Eight of the states which
have a general sales tax introduced it in the period 1965-69,
and another three states introduced it between 1960 and 1961.

Two other taxes displayed a similar expansion of coverage
during the 1960's: the individual income tax, which was
adopted by 11 states in the decade 1961-71, and the corporation net income tax, which was adopted by 9 states between
1963 and 1971. At the present, only 6 states do not have an
individual income tax--Florida, Nevada, Texas, Washington,
South Dakota and Wyoming; the last 5 states also do not have a
corporation income tax. Despite the expansion of coverage,
the latter tax is still a relatively minor component of total
tax revenues; the individual income tax, in contrast, is now a
major source of tax revenues, ranking third after general and
selective sales taxes. Over the same period, it should be

noted, property taxes declined in relation to total tax
revenues.

County Government Finances

As of 1972, county governments relied almost equally on
tax receipts and intergovernmental transfers for their
revenues. The bulk of the intergovernmental transfers comes
from the state governments, while the remainder--about 7% of
the intergovernmental transfers throughout the 1960's--comes
from the federal government and from local governments in
roughly equal proportions. Exhibit III-11 shows the composition of county government revenues, by source, in 1972 and
1962. The main features of this period were the growth in
intergovernmental revenues and a steady increase in the relative importance of receipts from current charges. Tax revenues
grew more slowly than total revenues, but, as Exhibit III-12
indicates, this was largely due to the heavy reliance on
property taxes; during the 1960's property tax receipts grew
relatively slowly at all levels of government. To compensate
for this, there was a significant increase in the use of
other taxes such as the general sales tax and the individual
income tax.* Taken together, non-property taxes doubled in
relative importance over the decade and, as of 1972, they
account for 14.4% of total tax revenues.

Municipality Finances

The Census of Governments draws a distinction, which is
not entirely unambiguous, between municipalities and townships. A municipality is defined as "a political subdivision
within which a municipal corporation has been established to
provide general local government for a specific population
concentration in a defined area."** The key features of
municipalities is that they are "created to serve specific
population concentrations." Townships, in contrast, "exist
to serve inhabitants of areas designated without regard to
population concentrations." The Census Bureau interprets this
definition of township to include the political subdivision of
"town" in New England, New York and Wisconsin, although it
concedes that this may be legally termed a municipal corporation and may have a governmental organization similar to that
of a municipality. Since townships (as defined by the Bureau)

_____

*There are no corporation income taxes at the county
government level.

**This and the following definitions are taken from the
1972 Census of Governments, Vol. 1, Governmental Organization,
pp. 2-3.

Exhibit III-11

County Government Revenue Sources and
Growth Trends: 1962 and 1972

Revenue Source	Amount ($ Million)		% Distribution		% Increase 1962-72
	1962	1972	1962	1972	
Intergovernmental Sources	$3,276	$ 9,956	37.7%	41.2%	+203.9%
Tax Revenues	4,149	10,076	47.7%	41.7%	+142.9%
Current Charges	824	2,697	9.5%	11.2%	+227.3%
Miscellaneous General Revenues	236	922	2.7%	3.8%	+290.7%
Utility Revenues	46	120	0.5%	0.5%	+160.9%
Liquor Store Revenue	71	120	0.8%	0.5%	+ 69.0%
Insurance Trust Revenue	92	283	1.1%	1.2%	+207.6%
Total Revenue	$8,694	$24,174	100.0%	100.0%	+178.1%

Note: Because of rounding, details may not add up to totals.

SOURCE: 1962 Census of Governments, Vol. IV, No. 2,
Finances of County Governments, Table 1.

1972 Census of Governments, Vol. IV, No. 3,
Finances of County Governments, Table 1.

Exhibit III-12

County Government Tax Collections in Detail and
Growth Trends: 1962 and 1972

	Amount ($ Million)		% Distribution		% Increase
Tax Source	1962	1972	1962	1972	1962-72
Property	$3,879	$ 8,625	93.5%	85.6%	+122.4%
General Sales & Gross Receipts		751		7.5%	
Selective Sales & Gross Receipts		148		1.5%	
Individual Income Tax	271	192	6.5%	1.9%	+435.4%
Motor Vehicle License		111		1.1%	
Other		249		2.5%	
Total	$4,149	$10,076	100.0%	100.0%	+142.9%

Note: Because of rounding, details may not add up to totals.

SOURCE: Same as for Exhibit III-11.

in these states, and also in Michigan, New Jersey and Pennsylvania, are generally vested with fairly broad powers of government and perform the same type of function as municipalities, we shall combine municipality and township finances in these states. In other states which have townships—Illinois, Indiana, Kansas, Minnesota, Missouri, Nebraska, North Dakota, Ohio and South Dakota*—they are located in predominantly rural areas and perform only a very limited set of functions; consequently, we shall not present any data on township finances in these states.

"Municipality," therefore, is defined in this report to include townships in New England, Michigan, New Jersey, New York, Pennsylvania, and Wisconsin. Exhibit III-13 presents an analysis of municipality revenue sources in 1962 and 1972. The most striking change over this period was the quadrupling of intergovernmental revenues from the federal and state

———————————

*There are also townships in one county of Washington, but these are very limited in their activities, and the Census does not provide financial data for these units.

Exhibit III-13

Municipality Revenue Sources and Growth Trends
1962 and 1972

Revenue Source	Amount ($ Million)		% Distribution		% Increase 1962-72
	1962	1972	1962	1972	
Intergovernmental Revenue	$ 3,003	$12,364	16.4%	26.9%	+311.7%
Tax Revenues	8,957	19,552	48.8%	42.5%	+118.3%
Current Charges	1,572	4,063	8.6%	8.8%	+158.5%
Misc. General Revenues	1,060	2,715	5.8%	5.9%	+156.1%
Utility Revenues	3,210	5,987	17.5%	13.0%	+ 86.5%
Liquor Store Revenues	77	103	0.4%	0.2%	+ 33.8%
Insurance Trust Revenues	460	1,190	2.5%	2.6%	+158.7%
Total Revenue	$18,340	$45,976	100.0%	100.0%	+150.8%

NOTE: Because of rounding, details may not add up to totals.

SOURCE: 1962 Census of Governments, Vol. 4, No. 3,
Finances of Municipalities and Township
Governments, Tables 1 and 18.

1972 Census of Governments, Vol. 4, No. 4,
Finances of Municipalities and Township
Governments, Tables 1 and 18.

governments; the latter, however, rose less rapidly than the
former: the proportion of intergovernmental revenues coming
from the state governments declined from 80% to 73%. Exhibit
III-14 shows the distribution of tax collections by municipalities in 1962 and 1972. By far, the largest increase in
revenues was registered by the municipal income tax, which is
now levied by approximately 3,817 localities in 10 states.
Most of these localities are in Pennsylvania (3,388 localities)
and Ohio (321 localities). However, local income taxes are
also to be found in Kentucky (34 localities), Indiana (31
localities), Maryland (24 localities, including 9 counties),
and Michigan (14 localities).* Most of these localities are

---

*In Indiana, Maryland, and Pennsylvania (with one excep-
tion), the tax is not levied on corporations. In the other
states, it is usually levied on corporation net income as well
as individual income.

Exhibit III-14

Municipality Tax Collections in Detail
and Growth Trends: 1962 and 1972

Tax Source	Amount ($ Million) 1962	1972	% Distribution 1962	1972	% Increase 1962-72
Property	$6,758	$13,308	75.4%	68.1%	+ 96.9%
General Sales & Gross Receipts	866	1,873	9.7%	9.6%	+116.3%
Selective Sales & Gross Receipts[a]	156	552	1.7%	2.8%	+253.8%
Public Utilities Receipts	281	766	3.1%	3.9%	+172.6%
Motor Vehicle Licenses	65	114	0.7%	0.6%	+ 75.4%
Income	259	1,881	2.9%	9.6%	+626.2%
Other[b]	571	1,058	6.4%	5.4%	+ 85.3%
Total	$8,957	$19,552	100.0%	100.0%	+118.3%

[a] Excludes Public Utilities Receipts.

[b] Includes non-property tax revenues of townships in 11 states.

NOTE: Because of rounding, details may not add to totals.

SOURCE: 1962 Census of Governments, Vol. 4, No. 3, Finances of Municipalities and Township Governments, Table 20.

1972 Census of Governments, Vol. 4, No. 4, Finances of Municipalities and Township Governments, Tables 7 and 18.

small in population, but there are also 49 cities with a 1970 population over 50,000; thirteen of these cities are listed among the 48 largest cities in the U.S. Some of the major cities which instituted an income tax during the 1960's are Baltimore, Detroit, Grand Rapids, Kansas City, New York City, Akron, and Cleveland.

Although income tax revenues accounted for only 9.6% of total tax revenues for all municipalities as a group, in the

cities which actually levy the tax it accounts for a significantly higher proportion of revenues. In the 13 largest
cities which have a local income tax, it provides an average
of about 40% of total tax revenues and, in several cases, it
provides over 50% of total tax revenues. To a large extent,
it has served as a substitute for the more traditional local
government tax sources. A Tax Foundation study in 1967 found
that "per capita property taxes are lower in the major cities
which impose income taxes than in those which do not. Moreover, in all but one case, per capita property taxes and per
capita total taxes have risen in the last decade by markedly
smaller percentages in the income tax cities."*

It appears that a similar pattern exists with respect to
the general sales tax. The share of sales tax receipts in
total municipality tax receipts stayed virtually constant over
the period 1962-1972, at 9.6%. However, an analysis of sales
tax revenues in the 14 largest cities with sales taxes in
1968 found that these revenues accounted for an average of
16.6% of total tax revenue.** The explanation of this
phenomenon is presumably that in larger cities the property
tax is relatively less adequate to finance municipal expenditure needs, and, therefore, there is a greater need to diversify revenue sources. This is borne out by the data in
Exhibits III-15 and III-16, which show the revenue structures
of municipalities in different size categories in 1962 and
1972. Over this period, the cities in the largest size
category reduced their dependency on property tax revenues
somewhat more than cities in other size categories. For
cities in the smallest size category, however, there was
relatively little change in the relative importance of the
property tax. The other noteworthy feature of this data is
that, while the intergovernmental revenues rose in proportion
to other revenues over the decade for cities of all sizes,
the increase was much more pronounced in cities of over
100,000. In 1962, intergovernmental revenues provided 22% of
total revenue for the largest municipalities and 16-18% of
total revenue for other municipalities; by 1972, these proportions had risen to 38% and 23-24%, respectively.

_____

*Tax Foundation, Inc., City Income Taxes, December 1967,
p. 34.

**Tax Foundation, Inc., State and Local Sales Taxes,
August 1970, Table 23.

58

Exhibit III-15

Amounts and Percentage Distributions of Selected Municipal Finance Items by Population-Size Groups: 1962

ITEM	All Municipalities	Municipalities with 1960 Population of				
		100,000 and over	99,999–25,000	24,999–10,000	9,999–5,000	less than 5,000
Amount of General Revenues/Capita	$112.88	$164.06	$98.28	$68.33	$58.64	$225.8
Percentage of General Revenues from:						
Intergovernmental Revenue	20.3%	22.1%	17.6%	16.0%	15.7%	17.4%
Tax Revenues	60.5%	60.9%	61.2%	57.6%	57.5%	59.6%
Current Charges	11.5%	10.5%	13.1%	16.6%	16.5%	2.2%
Special Assessments	2.0%	1.5%	2.8%	4.8%	3.6%	(a)
Amount of Tax Revenues per Capita	$68.27	$99.91	$60.34	$39.36	$33.60	$26.80
Percentage of Tax Revenues from Property Tax	73.2%	69.9%	81.7%	76.5%	75.7%	76.7%
Number of Municipalities	18,000	130	499	980	1,285	15,056
1960 Population (in millions)	116.3	51.0	25.6	15.1	9.1	15.9

(a) - Data not available

NOTE: This table temporarily excludes township finances in the eleven states where township finances are to be combined with municipality finances.

SOURCE: 1962 Census of Governments, Vol. IV, No. 3, Finances of Municipalities and Township Governments, Table 15.

Exhibit III-16

Amounts and Percentage Distributions of Selected Municipal

Finance Items by Population-Size Groups: 1972

ITEM	All Municipalities	Municipalities with 1970 Population of				
		100,000 and over	99,999-25,000	24,999-10,000	9,999-5,000	less than 5,000
Amount of General Revenues/Capita	$264.65	$415.81	$194.21	$143.75	$120.14	$97.52
Percentage of General Revenues from:						
Intergovernmental Revenue	32.9%	37.4%	24.4%	23.8%	23.1%	23.3%
Tax Revenues	48.6%	47.4%	53.5%	50.0%	48.1%	45.1%
Current Charges	11.1%	9.9%	14.1%	16.0%	17.2%	7.1%
Special Assessments	1.3%	0.7%	2.3%	3.4%	2.9%	1.3%
Amount of Tax Revenues per Capita	$128.62	$197.22	$104.43	$71.84	$57.73	$43.98
Percentage of Tax Revenues from Property Tax	64.3%	59.9%	74.8%	68.0%	68.4%	72.8%
Number of Municipalities	18,516	153	684	1,134	1,397	15,148
1970 Population (in millions)	132.3	56.5	31.8	17.7	9.8	16.4

NOTE: This table temporarily excludes township finances in the eleven states
where township finances are to be combined with municipality finances.

SOURCE: 1972 Census of Governments, Vol. IV, No. 4, Finances of Municipalities
and Township Governments, Table 16.

Special Sewer District Finances

In 1972, there were 2,042 special sewer districts in the U.S. located in 41 states. Of these, 1,411 were single-function districts and 631 were multiple-function districts, being engaged in the provision of water-supply as well as sewer services. The distribution of these districts among the states is shown in Exhibit III-17. This represents a considerable change from the situation in 1967--the only other year for which the Census provides data on special district finances--when there were 1,233 single-function sewer districts and 298 multiple-function sewerage and water supply districts.

Unfortunately, the available data on sewer district finances in both these years is relatively sketchy, as can be seen in Exhibit III-18. In 1967, total revenues of single-function sewer districts and multiple-function sewer and water supply districts amounted to $341.5 million; of this 34.7% came from tax revenues; comparable data on revenues from other sources, including charges, is not provided.* In 1972, total revenues amounted to $696.3 million, an increase of 104%; 20.5% of these revenues came from taxes.

However, in both years, the Census provides more detailed information on the finances of a number of larger special districts. In the 1967 census, 93 special districts were identifiable as single or multiple purpose sewer districts; in the 1972 census, there were 63 such special districts. The overlap between the two samples is, apparently, quite small: judging by the names of the districts, only 35 districts appear in both samples. This may be misleading because some districts may have changed their name between 1967 and 1972; it is also possible that there was consolidation of districts in some states. Beyond this, however, it appears that the Census Bureau changed its selection criteria between the two censuses.

The 1967 sample contains only 6% of the total number of single- and multiple-function sewer districts in the U.S., yet these districts account for 72.5% of the total revenues of all single- and multiple-function sewer districts in that year. The districts in the 1972 sample comprise only 3% of the total number of sewer districts. Exhibit III-19 shows the breakdown of revenues between property taxes and current charges in each sample. Although the samples are not strictly comparable for the reasons noted above, we can safely make certain inferences

---

*Data on charges by function, including sewerage, is available, but it is not classified by type of district--e.g., single-function versus multi-function district.

\multicolumn{3}{c}{Exhibit III-17}
\multicolumn{3}{c}{Number of Special Sewer Districts, by Function}
\multicolumn{3}{c}{and by States: 1972}
State	Sewerage	Sewerage & Water Supply
Arizona	8	
Arkansas	6	1
California	137	78
Colorado	146	93
Connecticut	4	3
Florida	4	3
Idaho	9	2
Illinois	128	5
Indiana	4	1
Iowa	12	
Kansas	3	2
Kentucky	32	3
Louisiana	31	
Maine	23	3
Maryland	1	8
Massachusetts	2	1
Michigan	6	8
Minnesota	2	
Missouri	6	
Montana	5	1
Nebraska	144	2
Nevada	6	4
New Hampshire	2	2
New Jersey	67	27
New Mexico		4
North Carolina	8	9
Ohio	3	4
Oklahoma	3	4
Oregon	40	2
Pensylvania	439	138
Rhode Island	2	
South Carolina	7	8
South Dakota	15	6
Tennessee	1	11
Texas	3	164
Utah	14	6
Virginia	4	
Washington	52	2
West Virginia	24	15
Wisconsin	3	
Wyoming	5	6
U.S. TOTAL	1,411	631

SOURCE: 1972 Census of Governments, Vol. IV, No. 2,
Finances of Special Districts, Table 15.

	Exhibit III-18

Numbers of Special Sewer Districts for Which Detailed
Information on Finances is Available by State
1967 and 1972

State	Number of Districts in Sample 1967	Number of Districts in Sample 1972	Number of Districts Common to Both Samples
Arkansas	1		
California	14	6	3
Colorado	5	2	1
Connecticut	1	1	
Idaho	1		
Illinois	12	7	5
Kentucky	1	1	1
Maine	1		
Maryland	2	2	1
Massachusetts		2	
Michigan	2	3	2
Minnesota	2	1	1
Missouri	1	1	1
Nevada	2		
New Jersey	15	13	6
North Carolina	2	1	1
Ohio	2	1	1
Oregon		1	
Pennsylvania	16	12	5
South Carolina	1	2	1
Utah	3		
Virginia	1	1	1
Washington	6	3	3
Wisconsin	2	3	2
Total	93	63	35

SOURCE: Computed from data in:

1967 Census of Governments, Vol. 4, No. 2,
Finances of Special Districts, Table 14.

1972 Census of Governments, Vol. 4, No. 2,
Finances of Special Districts, Table 14.

as to changes in revenue structures during the period 1967-1972. It seems clear that property tax revenues grew slowly in absolute terms and declined markedly in relative importance, while charge revenues grew in relative importance. We may speculate that the other major source of revenues was intergovernmental transfers, which probably grew significantly over the period.

Exhibit III-19

Finances of Selected Sewer Districts:
1967 and 1972

Item	Amount ($Million)		% Distribution	
	1967	1972	1967	1972
Property Tax Revenues	$102.5	$108.3	41.4%	23.9%
Current Charges	73.7	140.8	29.8%	31.1%
Other	71.4	203.7	28.8%	45.0%
Total Revenues	$247.6	$452.8	100.0%	100.0%

NOTE: 1967 data based on 93 sewer districts; 1972 data based on 63 sewer districts.

SOURCE: Same as for Exhibit III-18.

The Role of Sewer Charges

The previous sections have noted the increasing prominence of service charges as a source of state and local government revenues. Because of this trend, sewer service charges warrant closer analysis. There are two sources of data on the role of sewer service charges. One source are the Censuses of Government, the other source are special studies of sewer charges which have been made by such organizations as the American Water Works Association and the International City Management Association at various points in time. The data from each source will be examined in turn.

To judge their relative importance, sewer service charges should be related to the amount of expenditure on the provision of sewerage services by the units of government which collect the sewer charges. The relevant amount of local governments' expenditures on sewerage is difficult to assess for a number of reasons. It would not be reasonable to compare revenues from sewer charges in a given year with capital expenditures on sewerage in that year; first, because some of those expenditures might be financed from intergovernmental revenues, and second, because the government units probably will not attempt to pay for all of these expenditures out of

64

current revenues. , The only available data on intergovernmental transfers for sewerage pertains to the federal government — we have no information on state-local intergovernmental transfers for sewerage. Exhibit III-20 presents an analysis of sewer charge revenues in relation to sewerage expenditures corrected for the federal government's intergovernmental transfers for sewerage to all local government units combined.*

Exhibit III-20

Municipality & Special District Sewerage Charge Revenues and Expenditures and Federal Government Transfers for Sewerage: 1967 and 1972

Row	Item	Amounts ($ millions)		% Increase 1967-1972
		1967	1972	
a	Federal Government Intergovernmental Transfers for Sewerage*	$ 102	$ 913	+795.1%
	Local Government:			
b	Total Expenditures on Sewerage	1,563	2,856	+ 82.7%
c	Capital Outlays	995	1,822	+ 83.1%
d	Non-Capital Outlays	568	1,034	+ 82.0%
e	Revenues from Sewer Charges	544	1,151	+111.6%
f	(b) minus (a)	1,461	1,943	+ 33.0%
g	(c) minus (a)	893	909	+ 1.8%
	(c) as percent of (b)	63.7%	63.8%	
	(g) as percent of (f)	61.1%	46.8%	
	(e) as percent of (d)	95.8%	111.3%	

*Assumes that county governments receive 10% of federal aid to local governments for sewerage.

SOURCE: U.S. Department of Commerce, GF Series #5, Governmental Finances in 1971-72. Tables 8 and 9.

---

*This excludes county governments, for which there are no data on sewer charge revenues. We assume that county governments receive 10% of the federal government transfers for sewerage, on the basis of data presented in Table III-11.

65

Governments will finance their own share of the capital costs through long-term bonds to be repaid in future years. Therefore, the expenditure amount should be an annualized cost estimate, including debt service as well as operations and maintenance costs. However, there is a shortage of data concerning local governments' outstanding long-term debt for sewerage from which estimates of debt service costs could be constructed, with the single exception of special sewer districts.

Exhibit III-21 presents an analysis of special sewer district expenditures and charge revenues containing an estimate of debt service costs.* Finally, in Exhibit III-22 an analysis of municipality expenditures and charge revenues for sewerage containing neither of these corrections is presented. In Exhibit III-20 and III-22, non-capital outlays for sewerage include operations, maintenance and interest costs, but no annual debt repayments.

Exhibit III-20 shows that federal intergovernmental transfers for sewerage compensated for most of the growth in local governments' expenditures on sewerage over the period 1967-72. If it is assumed that all the federal aid went to subsidize capital outlays, it appears that local governments' capital outlays from their own sources remained virtually constant over that time. It should also be noted non-capital outlays rose at roughly the same pace as capital outlays, while sewer charge revenues rose significantly faster. In 1972, charge revenues exceeded non-capital outlays by 11.3%.

A comparison of Exhibit III-21 and III-22 shows that municipalities increased their use of sewer charges significantly more than special sewer districts over the same period. In 1972, municipalities' revenues from sewer charges significantly exceeded non-capital outlays, while special districts' charge revenues fell slightly short. Taking account of debt service charges as well as O&M costs, we find that, as of 1972, special districts' charge revenues covered only 48% of annual costs.

The other source of data on sewer charges are published and unpublished surveys of individual municipalities and special sewer districts. Included in the first category are

---

*We assume that debt service costs in 1967 amounted to 7.05% of the outstanding long-term debt which is the figure implied by the data in the 1961 MFOA. To allow for the secular growth in interest costs, we assume that 1972 debt service charges amount to 7.25% of the outstanding long-term debt in that year.

66

Exhibit III-21

Special Sewer District Charge Revenues and

Expenditures on Sewerage and Growth Trends:

1967 and 1972

Row	Item	Amounts ($ Millions)		% Increase 1967-1972
		1967	1972	
a	Total Expenditures on Sewerage	$ 344.7	$ 661.5	+ 91.9%
b	Capital Outlays	234.6	445.3	+ 89.8%
c	Non-Capital Outlays	110.1	216.2	+ 96.4%
d	Outstanding Long-Term Debt*	1,670.01	2,579.1	+ 54.4%
e	Estimated Annual Debt Repayments**	117.4	187.0	+ 59.3%
f	Estimated O&M and Debt Service Costs (c+e)	227.5	403.2	+ 77.2%
g	Revenues from Sewer Charges	84.3	193.6	+129.7%
	(g) as percent of (f)	37.1%	48.0%	

*Assumes that 70% of multiple-function water supply and sewerage districts' debt is for sewerage.

**Assumes that debt service charges amount to 7.05% of outstanding debt in 1967 and 7.25% of outstanding debt in 1972. See text for explanation.

SOURCE: 1967 Census of Governments, Vol. 4, No. 2, Finances of Special Districts, Tables 5, 7, & 13.

1972 Census of Governments, Vol. 4, No. 2, Finances of Special Districts, Tables 5, 7, & 13.

Exhibit III-22

Municipality Sewer Charges and Expenditures on

Sewerage and Growth Trends: 1962 and 1972

| Row | Item | Amounts ($Millions) | | % Increase |
		1962	1972	1962-72
a	Total Expenditures on Sewerage	$951.2	$2,194.4	+130.7%
b	Capital Outlays	637.9	1,377.1	+115.9%
c	Non-Capital Outlays	313.3	817.3	+160.9%
d	Revenues from Sewer Charges	305.4	957.5	+213.0%
	(d) as percent of (c)	97.7%	117.2%	

SOURCE: 1962 Census of Government, Vol. IV, No. 3, Finances of Municipalities and Township Governments, Tables 7, 10, 19.

1972 Census of Governments, Vo. IV, No. 4, Finances of Municipalities and Township Governments, Tables 8, 11, 19.

surveys by the APWA, MFOA and ICMA* as well as the FWPA** and
EPA.*** The results of this research will be summarized under
three headings: the growth in the use of sewer charges, the
structure of these charges, and their relationship to sewerage
costs.

The first American city to institute sewer service charges
was Brockton, Massachusetts in 1894. Before that time — and
for many years thereafter — it was common to levy a fixed
charge to cover connection costs but not to impose a fee proportional to the usage of sewer services. Very few other cities
introduced sewer charges until the 1930's, when the severe shortage of local government revenues from traditional sources combined with the flurry of public sewerage construction activity
to make them an attractive revenue source. Only 4% of the 162
U.S. cities surveyed by the MFOA had adopted sewer charges prior
to 1930, 20% instituted them during the 1940's, and 57% instituted them during the 1950's. As of 1961, the MFOA found that
63% of the U.S. cities which it surveyed had sewer service
charges. In 1953, the APWA study had concluded that about 30%
of the municipalities with populations over 5,000 had sewer
service charges. This figure may be a little too high. The
ICMA reports that in surveys of cities with populations over
10,000, less than 20% had sewer service charges in 1945, while
61% had sewer service charges in 1960.

In 1969, in a survey of more than 1,000 cities of all
sizes, the ICMA found that 86% of the cities which had a public
sewer system levied service charges. The prevalence of sewer
charges did not vary appreciably with population size--84% of the
cities with a population less than 5,000 had sewer charges, while
88% of the cities with a population over 500,000 had sewer charges.
The distribution of sewer charges by geographical region was more
significant; they were found in 72% of the cities in the Northeast,
84% of the North Central cities, 89% of the cities in the West, and
92% of the cities in the South.

The sewer service charges are computed in various ways.
Analytically, it is convenient to distinguish between (i) charges

---

*American Public Works Association, Sewage Service Charges
in Cities over 5,000, Special Report No. 18, Chicago, 1955;
Lennox L. Moak, Sewer Services Charges, Municipal Finance
Officers Association, Chicago, 1961; International City Management Association, Sewer Services and Charges, February 1970.

**FWPCA, The Cost of Clean Water and Its Economic Impact,
Vol. 3, Sewerage Charges, January 1969.

***Reported in EPA, The Economics of Clean Water: Vol. 1,
1972, pp. 142-5.

which are levied at a flat rate totally independent of water consumption or sewage flows,* (ii) charges which are based, at least in part, on the quantity of water consumed or discharged, and (iii) charges which are based on variables such as meter size, number or type of plumbing fixtures, or number of employees, which may be presumed to vary imperfectly with water consumption or sewage discharges. The MFOA survey in 1961 found that 12% of the charge systems were based on a flat rate, 77% were based at least partly on water use, and 11% were based on the number of fixtures or meter size, etc. Similarly the ICMA found in 1969 that 67% of the charge schedules for residential customers and 81% of the charge schedules for commercial and industrial customers were based, at least partially, on water or sewage flows. Charges based on a flat rate or on meter size or the number of fixtures, etc., were significantly more common in the West than in any other region; these charges were also slightly more common in cities with less than 5,000 population than in other cities, but, apart from this, the variation in charge method by city size was not very pronounced.

Some cities also levy a special sewer charge for industrial customers discharging wastes of greater than normal strength; this charge usually is based on the BOD and/or SS concentration of the wastewater discharges. Out of 1,160 cities surveyed by the ICMA in 1969, 233 had enacted ordinances providing for such charges, but only 57 were actually levying the charge. In 1972, the EPA reported the results of a survey showing 287 cities with charges based on the quality as well as volume of industrial wastes. This number is almost certainly too high. A telephone survey in 1973 of 120 cities reported by various authorities to have sewer charges based on wastewater concentration revealed that only 51 cities were actually applying the charges. About 40 cities indicated that they intended to levy such charges in the future. This number is bound to increase drastically as the industrial cost recovery provisions of P.L. 92-500 are enforced.**

One other feature of sewer rate structures is worth mentioning. Where charges are based on water consumption or sewage flows, they almost always follow a declining block pattern. The difference between the charges for the first block and the last block is often very large — a factor of two or three is not uncommon. Similar rate structures for electric utilities have recently come

---

*This includes charges to commercial and industrial users based on front footage.

**See Chapter I for a description of these provisions. At present, only four state governments have their own cost recovery requirements for communities which accept state funds for wastewater treatment facility projects.

under sharp criticism, and it is reasonable to ask whether the
same objection might be made against sewer rates. It lies beyond the scope of this report to examine this issue, but one can
say at a first glance that some of the arguments about the economic inefficiency (and possibly, inequity) of differential service charges (as opposed to connection charges) are likely to
apply to sewer rates.

This issue is related to the broader question of whether
at present sewer charges are adequate to cover sewerage costs.
Unfortunately, although this is a crucial question, very little
evidence is available from the surveys of individual cities.
The 1961 MFOA survey found that the total income from sewer
charges for 186 local government units was $117.6 million, while
the total debt service costs for 178 of these local governments
was $70.2 million and the total operating costs for 183 of them
was $69.5 million (the discrepancies are due to differences in
response rates for different items in the MFOA questionnaire).
This suggests that charge revenues cover about 80% of annual
costs. In a similar vein, an FWPCA report in 1969 concluded that
in most cities, sewer charge revenues were sufficient to cover
O&M costs alone, but not O&M costs combined with debt service
costs. The MFOA questionnaire contained an item asking about
the extent to which the jurisdiction considered its sewer activities to be self-supporting from sewer revenues. Nearly
three-quarters of the local governments claimed to be completely
self-supporting, a tenth claimed to be 75-99% self-supporting
and another tenth claimed to be 50-74% self-supporting. Finally,
in 1973-74, an unpublished survey of 42 cities levying special
charges on high strength industrial wastes found that officials
in 31 cities (74% of the sample) considered that these charges
were adequate to cover the additional costs of treating the industrial wastes.

Debt Financing

In most cases, the initial source of funding for publicly-
owned wastewater treatment facilities comes from some combination
of the following sources:

● intergovernmental transfers

● a capital improvement reserve fund

● short-term bonds, including anticipation notes

● long-term bonds.

Intergovernmental transfers and capital improvement reserve funds
are discussed elsewhere in this report. Short-term debt (defined
as debt with a maturity of less than one year) is often used by

71

local governments to obtain flexibility in planning their entry into the bond market for more permanent financing. Although the maturity of the debt is less than one year, it may be "rolled over" so as to provide a bridge of one or two years before permanent financing is sought.

By far the largest source of initial funding for capital projects is long-term debt, although it is hard to state precisely how important it is. The Joint Economic Committee study published in December 1966 concluded that new long-term debt accounted for approximately 50% of state and local government capital outlays for all functions over the period 1956-65.* Lennox Moak, in a similar analysis of local government capital outlays for all functions over the period 1958-67, concluded that the new borrowing accounted for approximately 66.7% of capital outlays.** As for sewerage expenditures, a FWPCA study, published in January 1968, presents data showing that state and local government bond sales over the period 1958-66 accounted for about 80% of construction contract awards for sanitary sewers and municipal waste treatment facilities.*** However, the EPA's 1973 report examined sewer bond sales by state and local governments in the period of 1961-70 and found that they amount to 57% of construction contracts, or 66.5% of the state and local share of construction contracts (i.e., with the federal government's grants deducted from the amount of construction contracts****). It should be noted that the difference between FWPCA and EPA estimates arises in part because they contain different figures for the annual amounts of construction contracts and sewer bonds in the years during which they overlap; no explanation is offered for the discrepancy.

The long-term debt instruments can be classified into three groups:

------

*U.S. Congress, Joint Economic Committee, State and Local Public Facility Needs and Financing, Volume 2, Supplement B, December 1966.

**Lennox L. Moak, "Financing Local Government Capital Improvements in the 1970's," Municipal Finance, February 1970.

***U.S. Department of the Interior, FWPCA, The Cost of Clean Water: Economic Impact on Affected Units of Government, January 1968, Table II. It should be noted that the text (p. 61) refers to the percentage of contract awards financed by sewer bonds as 90%, but this is not supported by the data in the table.

****U.S. Environmental Protection Agency, The Economics of Clean Water - 1973, December 1973, Table VII-3.

(a) General Obligation Bonds, secured by the full faith credit and taxing power of the issuing government.

(b) Revenue Bonds, payable only from the income from specific tax or revenue sources, but not secured by a pledge of the full faith and credit of the issuer.

(c) Special Assessment Bonds, payable from assessment against those who benefit from the facilities constructed with the proceeds of the bond sale (e.g., in the form of a tax on that property), but generally not secured by a pledge of the full faith and credit of the issuer.

The latter two types of debt are referred to collectively as nonguaranteed debt. General obligation (GO) bonds are the oldest and still the most common type of debt. Special assessment bonds became important during the urban land boom of the 1920's. At that time it was relatively easy for developers to create subdivisions and on the strength of this to issue special assessment bonds to be repaid by the future purchasers of the lots. The collapse of the land boom in the Depression led to many defaults on special assessment bonds, and they came to be seen as relatively unattractive debt instruments by investors; since the 1930's they have not been widely used to finance municipal improvements. In contrast, during the Depression, revenue bonds were considered a sound investment, since utility revenues were often a more certain source of income than taxes. Their popularity has continued to the present day, and they have greatly broadened in scope. Revenue bonds issued for utility-type operations account for only about half the total number of revenue bond offerings and less than half the dollar value of revenue bond sales. Overall, revenue and other non-guaranteed debt account for about 40% of the amount of long-term debt issued; for sewerage the share of revenue bonds is slightly lower, about 30%.*

This is a reflection of the fact-, noted in the previous section, that historically sewer services were not provided on a utility basis but were financed from general tax revenues. In the post-war period, this has changed with the growth of sewerage charges and the creation of many special sewer districts. For the most part, the choice between GO bonds and non-guaranteed debt hinges on three consideratons.

---

*This is based on data on sewer bond sales in the 1960's. More recent data are not available, although our survey suggests that the share of revenue bonds may have been closer to 40% in recent years.

73

- Constitutional and other statutory restrictions on the issuance of GO bonds, including limitations on the amount of such debt which may be outstanding, and requirements for voter approval of such debt.

- Interest rate differentials between GO and non-guaranteed debt which generally favor the safer GO bonds by about .25-.5 percentage points.

- The greater need for the issuer of non-guaranteed debt to prove their security to investors by demonstrating that revenues will be more than adequate to cover annual debt service costs. For sewer bonds a coverage rate (i.e., ratio of projected net revenues to debt service costs) in the range from 1.75 and 2.25 or higher is generally required.*

The general question of interest rate differentials in relation to the type of debt instrument and the type of issuing agency will be discussed further later in this chapter. At this point, let us examine three issues concerned with whether or not the need to use debt financing will put a constraint on the financing of wastewater treatment facilities required under P.L. 92-500: 1) the impact of statutory restrictions on the issuance of guaranteed debt, 2) the effect of cyclical interest rate fluctuations on the timing of investment, and 3) the special problems of small municipalities in the bond market.

State Constitutional and Statutory Restrictions on Local Government Debt: These restrictions take the following forms: (a) limits on the amount of outstanding local government debt in relation to the property tax base; (b) limits on the rates of certain taxes (especially property taxes) which can be levied for debt service or other purposes, or limits on the size of annual tax increases for this purpose; (c) requirements for specific voter approval of proposed bond issues, (d) limits on interest rates which can be paid on local government bonds. The first three restrictions apply mainly to GO bonds. The last type of restriction generally applies to revenue as well as GO bonds and its effect will be discussed separately below. In practice, the distinctions among the first three types of restrictions are somewhat artificial, since in some cases the limits on the amount of outstanding debt or on tax rates can be waived if approved by a referendum. Moreover, in most states the legislation containing the debt restrictions is extremely complex and laden with

---

*Paul D. Speer, "Planning for Revenue Bond Financing," MFOA, Special Bulletin 1968A, February 16, 1968.

74

numerous exceptions and special provisions regarding debt issued by special government agencies and for specific purposes.* Consequently, it is hard to quantify in a single measure the extent of the debt restrictions in each state.

Exhibit III-23 presents a brief summary of existing restrictions on local government GO debt. The first column shows whether a referendum is generally required for GO debt. The second column shows whether there is a limit on the amount of outstanding debt as a proportion of the property tax base (assessed at market value, locally established assessed value, or state equalized assessed value). The third column indicates the interaction between the two sets of restrictions — specifically, it shows whether there is still some restriction on the amount of GO debt even after voter approval has been obtained. The last column is intended to measure the overall severity of the debt restrictions. There are 22 states in which these restrictions are deemed to be relatively mild, because the state constitutions are entirely or nearly free of limitations on local government debt (such states are denoted by "0") or for other reasons.** It should be emphasized that this list is only approximately accurate and suffers from numerous deficiencies. For example, our sources have provided only summary information pertaining to municipalities and counties — they do not cover special sewer districts, for which there are frequently special provisions. Moreover, there are sometimes specific exceptions for sewerage expenditures,and these are not generally recorded

---

*We may quote two examples of the complexity of state legislation dealing with local government debt, taken from an ACIR report. The report cites a recent study of state supervision of municipal debt in Kansas which identified 433 separate statutory grants of debt raising power to local governments in that state. It also mentions an earlier study of local finance in Illinois which noted 167 separate provisions for the issuance of GO bonds by local government units, requiring 15 pages of small print for tabulation. The ACIR report comments "...under such circumstances, no single percentage might correctly be said to represent 'the' limit on general obligation debt that could be incurred by a particular class of local governments or even by a particular local government." ACIR, State Constitutional and Statutory Restrictions on Local Government Debt, Report A-10, September 1961 (p. 29).

**This tabulation represents our own assessment, based on data contained in the ACIR report and other sources mentioned at the foot of the Exhibit.

Exhibit III-23

Constitutional and Statutory Limitations on Local Government
Sewerage General Obligation Debt

State	Referendum Requirement (1)	Debt Limit (2)	Debt Limit With Voter Approval (3)	Severity of Restrictions (4)
Alabama	M	20(LAV)	1	1
Alaska	M	-	0	0
Arizona	M	19(EAV)	1	1
Arkansas	N	t	1	1
California	2/3	15(LAV)	0	0
Colorado	M	V	1	1
Connecticut	-	V	NA	0
Delaware	M	V	0	0
Florida	M	10(LAV)	1	1
Georgia	M	7(LAV)	1	1
Hawaii	-	15(MV)	NA	0
Idaho	2/3	15(MV)	1	1
Illinois	M	5(EAV)	1	1
Indiana	-	2(LAV)	NA	1
Iowa	2/3	5(MV)	1	1
Kansas	M	V	1	0
Kentucky	2/3	3-10(MV)	1	1
Louisiana	M	10(LAV)	1	1
Maine	M	7.5(LAV)	1	0
Maryland	M	-	0	0
Massachusetts	-	V	NA	0
Michigan	M	10(EAV)	1	1
Minnesota	M	V	0	0
Mississippi	-	15(LAV)	NA	0
Missouri	2/3	20(EAV)	1	1
Montana	M	10(EAV)	1	1
Nebraska	M	-	0	0
Nevada	M	10(LAV)	1	0
New Hampshire	-	1.75(EAV)	NA	0
New Jersey	-	3.5(EAV)	NA	0
New Mexico	M	4(LAV)	1	1
New York	-	7-10(MV)	NA	0
North Carolina	M	8(LAV)	0	0
North Dakota	3/5	8(EAV)	1	1
Ohio	M	10(LAV)	1	0
Oklahoma	3/5	5(LAV)	1	1
Oregon	M	3(MV)	1	1
Pennsylvania	M	15(LAV)	1	1
Rhode Island	M	3(LAV)	0	0
South Carolina	M	8(LAV)	1	1

See Notes on Next Page.

76

Exhibit III-23 (Cont'd) Constitutional and Statutory Limitations on Local Government Sewerage General Obligation Debt				
State	Referendum Requirement (1)	Debt Limit (2)	Debt Limit With Voter Approval (3)	Severity of Restrictions (4)
South Dakota	3/5	15(EAV)	1	1
Tennessee	-	-	NA	0
Texas	M	-	1	1
Utah	M	4(MV)	1	1
Vermont	M	10(LAV)	0	0
Virginia	M	18(LAV)	1	0
Washington	-	5(LAV)	NA	1
West Virginia	3/5	5(LAV)	1	1
Wisconsin	-	5(EAV)	NA	1
Wyoming	M	6(EAV)	1	1

NOTES: Col. 1: " - " signifies no general requirement for a referendum. M signifies simple majority approval required. If other than majority approval is required, the appropriate fraction is recorded.

Col. 2: " - " signifies that there is no general limit on amount of debt that may be incurred. Otherwise the limit is indicated as a percentage of the property tax base assessed at market value (MV), locally established assessed value (LAV), or state equalized assessed value (EAV). In general, we have recorded the limits applicable to municipalities. "V" indicates that various rates apply to different classes of municipalities. "t" signifies that there is no limit on the amount of debt as a proportion of the tax base, but there is a limit on the tax rate which may be levied for debt service.

Col. 3: 0 = there is no limit on the amount of debt if suitable voter approval is obtained.
1 = even with voter approval, there are still limits in some circumstances.

Col. 4: 0 = the restrictions on local government issuance of debt are relatively mild.
1 = the restrictions are not mild.

SOURCES: Columns 1 and 2: ACIR, Federal-State-Local Finances, Report M-79, February 1974 (Tables 94 and 93).

Column 3: Thomas F. Pogue, "The Effect of Debt Limits: Some New Evidence," National Tax Journal, March 1970 (Table 1, column 5).

Column 4: Based on information presented in ACIR Report A-10, Thomas F. Pogue (op.cit), and A.J. Heins, Constitutional Restrictions Against State Debt, University of Wisconsin Press, 1963.

77

in the summary tabulation of state debt restrictions.* Hence, some caution must be exercised in making inferences as to the constraints on the capacity of local governments to finance wastewater treatment facility construction.

The impact of these restrictions on local government finance and expenditure patterns is not entirely clear. There is general agreement among those who have studied the problem that these restrictions are excessively rigid and frequently undesirable, but there is relatively little tangible evidence of their harmful effect. There can be no doubt that they are partly responsible for the growth in non-guaranteed debt in the post-war period and the concomitant phenomenon of prolific creation of special purpose agencies endowed with their own borrowing power. To the extent that revenue bonds require higher interest payments (or have higher administrative costs), this may be considered an example of the inefficiency of debt restrictions. However, it should be noted that special districts offer certain administrative advantages, and revenue bond financing, too, has certain advantages and may even be justified in terms of the benefit principle of taxation. Therefore, it is possible that these twin phenomena would have become so prominent even without the stimulus of debt limits; the phenomena may also not be an entirely undesirable outcome.

The more important issue is whether or not debt "limits" have prevented local governments from undertaking desirable expenditures and, specifically, whether or not they are likely to impede the wastewater treatment facilities construction program required by P.L. 92-500. Historically, there is some evidence that debt limits may have restrained the total volume of local government borrowing in those states where the limits have been relatively strongly enforced. This was the conclusion of the ACIR study, which compared outstanding long-term debt per $1,000 personal income in the states with and without specific constitutional restrictions on local government debt. A similar analysis was performed by Ratchford using both per capita debt and debt as a percentage of personal income, and a similar con-

---

*In some cases, the source on which column 2 is based indicated that a special percentage limit was generally applicable to GO debt issued for sewerage projects. In all such cases, this is the figure which we have recorded in the table.

clusion was reached.* However, Heins** reviewed Ratchford's data, correcting it to take account of certain special circumstances, and found that the differentials in the volume of debt were quite narrow. Moreover, Pogue+ found that although per capita debt levels were somewhat lower in states with more severe debt limits, there was not a statistically significant relationship between an index of debt limit severity and the debt levels, and the overall impact on local government borrowing (i.e., the sum of the effects on expenditure levels and on non-debt revenues) was approximately zero.

In addition to these somewhat negative findings, it should be noted that local governments have the power to evade debt limits by resorting to non-guaranteed debt. Further, state legislations generally have the power to circumvent the impact of debt limits if the expenditures which are to be financed are considered sufficiently important; in most states there is a vast body of legislation conferring special borrowing powers for various purposes. For these reasons, it does not appear likely that state debt limits are likely to present a significant obstacle to the fulfillment of objectives of P.L. 92-500. It is far more likely that the relevant constraint will not be the legal borrowing power of affected units of government, but rather their actual fiscal capacity to service the debt incurred. This raises issues about the future strength of the bond market and the buoyancy of local government revenues which lie beyond the scope of this study.

The Impact of Interest Rate Fluctuation on the Timing of Sewer Investments: Cyclical fluctuations in interest rates may affect the timing of investments in public wastewater treatment facilities either because of statutory limitations on interest rates for local government bonds or because of local governments' natural reluctance to assume the burden of high interest payments on relatively long-term capital projects. The statutory restrictions on interest rates are summarized in Exhibit III-24. It should be noted that the underlying legislation is extremely complex and contains many special provisions for debt issued for certain purposes or by certain governmental units. However, there is considerable reason to doubt that these restrictions have significantly affected the timing or volume of local government investment, especially for sewerage. First, state legislatures

---

*B.V. Ratchford, "State and Local Debt Limitations," National Tax Association Proceedings, 1958.

**A.J. Heins, op.cit.

+T.F. Pogue, op.cit.

79

EXHIBIT III-24
Maximum Allowed Interest Rates on Local Government Bonds

| State | Long-Term Debt | | Short-term |
	GO Bonds	Revenue Bonds	Notes
Alabama	8%	8%	V
Alaska	0	0	0
Arizona	0	0	0
Arkansas	6%	8%	V
California	7%	V	0
Colorado	0	0	0
Connecticut	0	0	0
Delaware	V	V	V
Florida	7½%	7½%	7½%
Georgia	0	9%	0
Hawaii	7%	7%	7%
Idaho	0	0	0
Illinois	7%	7%	V
Indiana	0	0	0
Iowa	7%	7%	7%
Kansas	7%	8%	0
Kentucky	0	0	0
Louisiana	6%	6%	6%
Maine	0	0	0
Maryland	V	V	V
Massachusetts	0	0	0
Michigan	8%	8%	8%
Minnesota	7%	0	0
Mississippi	6%	6%	0
Missouri	8%	8%	0
Montana	7%	9%	0
Nebraska	0	0	0
Nevada	8%	8%	8%
New Hampshire	0	0	0
New Jersey	0	0	0
New Mexico	8%	8%	0
New York	0	0	0
North Carolina	0	0	0
North Dakota	0	0	0
Ohio	8%	0	8%
Oklahoma	7½%	7½%	0
Oregon	8%	0	0
Pennsylvania	6%	0	6%
Rhode Island	0	0	0
South Carolina	7%	7%	7%
South Dakota	8%	0	0
Tennessee	10%	10%	10%
Texas	10%	10%	0
Utah	8%	8%	8%
Vermont	0	0	0
Virginia	0	0	0
Washington	0	0	0
West Virginia	8%	7%	0
Wisconsin	8%	8%	7%
Wyoming	0	0	0

NOTE: 0 = no restrictions; V = various restrictions

Source: The Bond Buyer, Municipal Finance Statistics, Vol. 11
          May 1973.

have proved willing to relax interest rate limits at times of
tight money conditions in bond markets; this occured both during
the 1969-1970 "crunch" and over the last eighteen months when
interest rates have reached record highs. Second, local governments may switch to short-term debt which is generally subject
to fewer restrictions than long-term debt; this also provides
an opportunity for delaying the permanent financing of capital
projects until more favorable long-term interest rates can be
obtained. Thus, short-term borrowing by local governments
trebled during the 1969-70 crunch, and there has been a similar
increase in short-term borrowing in the recent tight conditions.

All this should not be taken as proof that high interest
rates have absolutely no effect on local government capital expenditures. A considerable portion of the bond issues displaced
in the 1969-70 crunch, but not all, was subsequently reoffered
after the tight money conditions had eased. Kurtz suggests that
about 10-20% of the deferred bond issues were ultimately aban-
doned.* However, this estimate applies to local government
borrowing in aggregate. The record suggests that local government capital outlays for sewerage are among the least sensitive
to interest rate fluctuations. Thus, Frank Morriss, in a study
of bond issues, interest rates and contract awards over the
period 1952-59, found that most outlays were somewhat sensitive
to interest rate fluctuations, with two significant exceptions
— contract awards for educational and water/sewer facilities.**
In the Brookings quarterly model of the U.S. economy, the best
equations for state and local government capital outlays did not
contain an interest rate variable.*** In view of these findings,
we doubt that future interest rate fluctuations will significantly affect the timing of capital outlays for public wastewater
treatment facilities required by P.L. 92-500.

The Problems of Small Municipalities: Small municipalities
are frequently at a disadvantage in the bond market, for various
reasons. The Joint Economic Committee (JEC) study provides a
very thorough analysis of the problem. It lists three specific
factors which tend to discriminate against small borrowers:

---

*H.W. Kurtz, "Impact of Interest Rate Limitations," Muni-
cipal Finance, August 1970.

**Frank Morriss, "Impact of Monetary Policy on State and
Local Governments," Journal of Finance, May 1960.

***J. Duesenberry; G. Fromm; L. Klein; and E. Kuh (eds.),
Brookings Quarterly Econometric Model of the United States,
Rand McNally & Co., Chicago, 1965.

First, small municipalites market bond issues at infre-
quent intervals, and these issues usually involve only a
limited number of bonds of relatively small total dollar
amounts. However, overhead costs incurred in marketing
an issue of small dollar amount is not proportionally less
than the cost incurred in marketing a sizable issue. As
a consequence, market costs per bond are higher for small
issues, because the "spread" is greater for a small issue
than it is for a large issue. Major bond buyers, such as
insurance companies and commercial banks, usually prefer
to purchase bond issues that are large in total dollar
amounts because larger issues are generally easier to
trade. Thus, bond issues of small municipalities are
relatively more costly to market, and less attractive to
investors, than are the issues of large municipalities.
Second, large municipalities generally can provide quickly and accurately the detailed financial information
needed by bond dealers and buyers for an analysis of investment possibilities. Third, small municipalities
usually cannot afford to employ the experienced legal and
financial advisers necessary to guide the bond issue
through the intricacies of the bond market smoothly and
effectively.*

In addition, the bonds of small communities are more likely to be unrated than those of larger communities — which causes
them to bear higher interest rates — both because the bond
rating companies refuse the bonds of governmental units with less
than a minimum amount of outstanding debt,** and also because
the costs of obtaining a bond rating are proportionally higher
for smaller than for larger communities.*** There is also some
evidence that, even when the bonds of small communities have the
same rating as the bonds of larger communities, they still carry
higher interest rates. The evidence comes mainly from the JEC
study, which compared the net interest costs of bonds issued by

---

*U.S. Congress, Joint Economic Committee, op.cit.,
Vol. II, pp. 249-250.

**According to the JEC study, Moody's does not rate bonds
unless the governmental unit has at least $600,000 of outstand-
ing debt; the minimum outstanding debt required by Standard &
Poor's is $1 million.

***Thus, Moody's bases its fees for rating GO bond issues on
the population size of the issuing community; the fee (in 1971)
for a community with less than 50,000 population was $600 while
the fee for a community with over 1 million population was only
$1350. J.F. Clark, "Observations Concerning the Rating of Muni-
cipal Bonds and Credits," MFOA Special Bulletin, October 1, 1971.

communities with a population under 10,000 and communities with
a population between 10,000 and 250,000 over the period 1961-65.
Using the data presented in the JEC study, we have calculated a
weighted average of the net interest costs paid by small and
larger communities for bonds in the same rating class with a
maturity of 10-19 years — the time span generally used in financing sewer projects.* For "A" rated bonds, the average interest cost of smaller communities' bonds was 7.6 basis points
higher; for "B" rated bonds, the smaller communities' net interest cost was 16.4 basis points higher; and for unrated bonds,
the smaller communities' net interest cost was 112.6 basis points
higher.

The data are not entirely conclusive because the rating
categories are aggregate: the A category contains bonds rated
Aaa, Aa and A, while the B category contains bonds rated Baa,
Ba and B. Thus, it is still possible that the interest differentials are partly due to smaller communities' receiving lower
ratings within each bond category. This hypothesis is consistent
with the distributions of the JEC data across rating categories;
these distributions for small and larger communities are shown
in Exhibit III-25. It is clear that the small communities'
bonds are more likely to be unrated and less likely to have A
rating than those of large communities. The JEC report does not
indicate whether these data are a representative sample of bonds
issued by small and larger communities over the period. If so,
we can apply a formal statistical test of the hypothesis that
small communities' bonds have the same rating as those of larger
communities; this hypothesis is decisively rejected by the data.**
However, it appears that this conclusion is due entirely to the
greater predominance of unrated small communities. If we drop
the third column in Exhibit III-25 and use a chi-square test on
the remaining 2x2 classification, we find that we cannot reject
the hypothesis that small and larger communities are equally
likely to have A- and B-rated bonds, given that their bonds are
rated.***

---

*The data come from the same source as Exhibit III-25
below. The weights are the total number of bond issues by small
and larger communities in the given category in each of the five
years.

**There are 2 degrees of freedom and the chi-squared statis-
tic is 2548.

***There is 1 degree of freedom and the computed chi-squared
statistic is 2.24.

83

Exhibit III-25

The Ratings of Small and Larger Communities' Bond Issues:
1961-66

Community Size	Type of Rating		
	A	B	C
Small	504	525	1200
Large	3012	3469	459

SOURCE: Computed from the data in JEC, State and Local
Public Facility Needs and Financing, Vol. II,
p. 249, Table 1.

Although it seems clear that small communities do face
special hardships in the use of debt finance, it is unlikely that
this fact by itself will prevent them from fulfilling the requirements of P.L. 92-500. However, it does raise the question
of whether steps could be taken to ease their access to the bond
market. The answer to this question lies beyond the scope of
this report, but it is worth noting some of the solutions which
have been proposed. The main recommendation of the JEC report
was that state governments provide greater assistance to local
governments in the preparation and administration of bond issues,
on the lines of existing programs in Michigan, North Carolina,
Tennessee, Virginia and certain other states. A more innovative
solution is provided by the Vermont Municipal Bond Bank. This is
a state agency which assembles a group of local bond issues, sells
an issue of its own equal to the total amount of the local issues
(plus something for a reserve fund) and, with the proceeds, purchases the local bonds. The advantage to the localities is that
they effectively acquire the lower interest rates available to
state government bonds; at the same time they avoid some of the
high overhead costs incurred when small issues are brought to the
market. The scheme appears to be working very successfully and
has attracted the attention of many other states. It is also
possible that similar results could be obtained through the
operations of an agency such as the New York State Pure Waters
Authority, which is legally enpowered to contract with municipalities to construct wastewater facilities on their behalf and
to make loans to them for this purpose.

INCIDENCE OF VARIOUS METHODS OF FINANCING
WATER POLLUTION ABATEMENT

The costs of water pollution abatement may be financed
initially through debt instruments or other intermediate tools
such as the environmental trust fund. The ultimate financing,
however, must involve some combination of tax/user charge increases and expenditure substitution. The burden or distribution
of these costs across the population is defined as the "incidence" of the program costs. In this section, estimates are made
of the incidence of each of the financing techniques described
earlier.

Incidence of Borrowing

In a simple model, the use of bonds to finance the costs of
the Act introduces no new problems into the analysis of incidence.
In particular, issuing bonds is simply an intermediate step in
the financing process; eventually the cost of these bonds must be
repaid using one of the tax increase or expenditure substitution
techniques discussed earlier. The incidence of borrowing costs,
then, is simply equal to the incidence of techniques used to finance the bonds.

Within the context of this project, however, there is one
further wrinkle introduced by the use of local and state borrowing. There are differences among municipalities and states on
interest rates they must offer on their bonds; these rates reflect
bond buyers' judgements on the fiscal strength of different areas
which vary systematically by city size, median income of the area
population, and other factors. These interest rate differences
contribute to real differences across areas in the cost of con-
structing equivalent waste treatment facilities.

The environmental trust fund is another immediate tool which
may be used for financing water pollution control expenditures,
analogous to debt financing. Its incidence pattern depends upon
the method of repayment.

Tax Incidence

There are four major, relatively current studies which provide
estimates of the incidence of the full package of federal, state
and local taxes: Musgrave (1973), Herriott & Miller (1969),
Gillespie (1965), and the National Tax Foundation (1967). The
bulk of the distributional work presented in this report is predi-
cated on the Musgrave estimates, which were judged to be the most
reliable. In this section, therefore, only Musgrave's estimates
are presented.

Despite differences among the tax incidence studies, each
of the major studies follows a common four step procedure in estimating the distribution of the burden of taxation among income
groups.

1. The absolute amount of taxes, by type, collected in the
    test year(s) is determined. Typically this information
    is taken directly from national income accounts.

2. The proportion of each particular tax which is ultimate-
    ly borne by owners of particular factors of production
    versus consumers is estimated. This task requires more
    than a simple identification of the initial recipient of
    a tax. Typically, the imposition of a tax will stimulate
    some adjustment in the economy; one of the by-products of
    this adjustment may be a shifting of at least part of the
    tax to another group in the population. In incidence
    studies, the major concern is with identifying the ultimate
    recipient of the tax, once expected adjustments have oc-
    curred. This task is known as analysis of "Shifting."

The allocation of the tax burden among factors of production
and consumers derived in (2) above is transformed into an allocation of taxes among income groups. This requires investigators
to:

3. Define and calculate an appropriate income measure; on
    the basis of this measure the population is then sub-
    divided into groups.

4. The relationship between income and (a) the ownership of
    particular factors of production; and (b) consumption ex-
    penditures is estimated.

5. The aggregate taxes determined in step (1) are then dis-
    tributed among income groups. Typically, incidence
    studies estimate both the percentage of the total tax
    burden borne by different income groups, and the ratio
    of tax burden to income for each income group.

In calculating the ultimate incidence of different taxes,
different shifting assumptions can be made. The assumptions
used by Musgrave are summarized in Exhibit III-26.

Income Taxes: Musgrave assumes that the ultimate burden of the
income tax is distributed in accordance with the distribution
of the initial tax payments. This assumes that people do not
generally reduce their work hours in response to changes in this
tax (i.e., supply is inelastic).

Exhibit III-26	
Shifting Assumptions	
(Musgrave)	
TAX	**ASSUMPTIONS**
1. Income Tax	On the individual
2. Gift, Estate & Death Taxes	On total capital income over $25,000
3. Excise Taxes	On the consumer of the product
4. General Sales Tax	On general consumers
5. Corporate Profits Tax	a. Split between consumers & owners of capital (1/3, 2/3)
	b. Split between consumers, owners of capital & labor (1/4, 1/2, 1/4)
	c. All on capital
	d. All on consumption
6. Property Tax--Owner-occupied Residential	On home owners
Rented Residential	a. On renters
	b. Split between owners & renters
	c. On capital
Business Property	a. On capital
	b. Split between capital and consumers (1/2, 1/2)
	c. On consumers
7. Motor Vehicle License Tax	a. On general consumer
	b. On auto consumer
8. Customs Tax	On general consumers
9. Social Insurance Taxes-- Social Security	a. Split between labor & consumers (1/2, 1/2)
	b. On labor
Unemployment Taxes	a. Split between labor & consumer
	b. On labor
Retirement Taxes	On labor

Gift, Estate & Death Taxes: The initial recipient of a gift, estate or death tax is the donor; it is assumed that this tax cannot
be shifted. Hence, Musgrave distributes this tax in proportion to
the wealth held by particular income groups. In addition, low income groups (income <$25,000) are exempted from participation in
this tax. Presumably, this exclusion is done to reflect the legal
exemption of small estates from taxation.

Excise Taxes: Specific Taxes on Particular Goods: Suppose a tax
is imposed on the sale of one good only. Who ultimately pays this
tax? The effect of a selective sales tax will be to shift up the
supply curve of the firm by the size of the tax. If demand were
completely unresponsive to price (inelastic), the tax would result
in an equal increase in price; that is, consumers of the taxed good
would bear the full cost of the tax. It is rare, however, for demand to be perfectly inelastic; thus, in all likelihood, price of
the taxed good will rise by less than the tax. Who bears the remainder of the tax burden? In the short run, the producers of the
taxed good bear the remaining tax burden. In the long run, however,
after full adjustments in the economy occurs (i.e., factors of production rearrange themselves, demand for other substitute products
increases, and so on), we would expect the producers' burden of the
tax to be shifted to consumers of other goods in the economy.*
The magnitude of the burden of a tax on a particular good borne by
consumers of other goods will depend on the elasticity of substitution between the two goods.

In Musgrave's study, the burden of specific excise taxes is
placed on the consumers of the taxed goods. In the case of gasoline
and automobile excise taxes, Musgrave shifts one-third of the burden
to general consumers, using the long-run argument sketched above.

General Sales Tax: A general sales tax is equivalent to a tax on
consumption income. Musgrave attributes the full sales tax to
current consumption, and allocates it according to current consumption patterns.

The Corporate Income Tax: The corporate income tax represents a
relatively large source of tax revenues. Moreover, the incidence
of this tax is not yet well understood.

The corporate income tax is a tax on capital income origina-
ting in the corporate sector. In the short run, capital is fixed.
Hence, the full burden of the tax resides with owners of corporate
capital. Our real concern, however, is with ultimate burden, once
longer run adjustments in the economy have occured. Since capital

_____

    *Richard A. Musgrave & Peggy B. Musgrave, Public Finance in
Theory and Practice, McGraw Hill, 1973, New York.

is not fixed in the long run, some shifting of the tax is possible.

In a perfectly competitive, frictionless system the tax on corporate capital would, in the long run, induce capital to move out
of the corporate sector into the non-taxed sector. Indeed, this flow
would stop only when the net return on capital in all sectors was
equalized. In other words, the movement of capital in response to
the tax would spread the burden of the tax across all owners in capital. The analysis becomes somewhat more complicated when one recog-
nizes that:

- imperfections in capital markets mights prevent capital
  flows, and thus mitigate the tendency of the system to
  spread the tax;

- the shift in capital between sectors will shift production in the two sectors, thus changing relative demand
  for capital and labor and the rates of returns;

- the shift in production may change relative prices of
  the final good.

Once all of the complications and long-run adjustments are considered, three potential recipients of the corporate income tax burden can be identified: consumers, owners of capital, and labor.
Musgrave indicates his uncertainty about the burden distribution by
making four different assumptions about the burden of the corporate
income as:

- 1/3 to consumers, 2/3 to owners of capital

- 1/2 to owners of capital, 1/4 to consumers, 1/4 to labor

- total to owners of capital; and

- total tax to consumers.

The Property Tax: The incidence of the property tax, the major component of local tax revenues, is also not well understood, and hence
admits of a number of alternative shifting procedures.

Musgrave assumes that the tax on owner-occupied housing is not
shifted but is absorbed by the home-owner. Thus, this portion of
the property tax is allocated according to the housing expenditures
in each income group. Musgrave also assumes that property taxes
accrue to renters, in proportion to rental expenditures.

Property taxes on business property must also be allocated.
The general scenario is that capital flees the taxed business, thus
increasing prices (i.e., shifting part of the tax to consumers), and
depressing the return to capital in other non-taxed industries (thus,

spreading the tax among all owners of capital). The relative proportion of the tax appropriately attributed to consumers versus
capital is a matter of speculation: Musgrave assumes an allocation
of half to consumers, half to capital. Similar allocations are
used for the other taxes.

Using the methodology described above, Musgrave developed some
estimates of the percentage of particular taxes paid by specific income groups. These estimates are presented in Exhibits III-27 and
III-28 which follow.

The Incidence of User Charges

As described earlier in this chapter, user charges represent
an increasingly important method of financing the local share of
water pollution control expenditures; they are particularly important for financing the O&M charges which are borne almost exclusively
by local governments. However, the four incidence studies described
above present no estimates on the distribution of the user charge
burden. Therefore, during the course of this research, an estimate
of the distribution of the user charge burden was made using the
methodology described above for the other taxes.

User charges are levied on industrial, commercial, and residential customers. Industrial and commercial charges may be considered
similar to the general sales tax, with the charge rates varying for
particular categories of goods and services, depending on the effluent
characteristics of their production processes. Industrial and commercial user charges are then assumed to be borne entirely by consumers in the form of increased prices. The relevant evidence and
calculations are presented in Chapter 4 of this report, along with
other price increases.

Charges levied on residential users do not "shift," they are
borne entirely by residential users. Residential user charges
may be grouped into three major categories:

● Flat charges: these are per capita or more usually per
  family charges.

● Volume based charges: these are levied on the basis of
  metered water use or some imperfect substitute, such as
  the number of plumbing fixtures, pipe size, and so on.

● Charges collected from the property tax: the incidence
  of these charges is analogous to the incidence of the
  property tax; however, since they represent simply a
  part of another fiscal instrument, the property tax,
  they are not properly considered user charges.

Flat charges may be distributed in line with population/income

90

Exhibit III-27

Distribution of Federal Taxes by Income Group:

Musgrave

Income Category	Personal Income Tax Percentage of Tax Paid by Each Group: Base =$76.3b	Corporate Income Tax Percentage of Tax Paid by Each Group: Base =$38.3b	Excise Tax Percentage of Tax Paid by Each Group Base =$15.1b	Gift & Estate Tax Percentage of Tax Paid by Each Group: Base =$3.2b	Customs Tax Percentage of Tax Paid by Each Group Base =$2.3b	Social Insurance Tax Percentage of Tax Paid by Each Group Base =$40.5b
$ 0-4,000	0.8%	4.1%	4.3%	0%	5.7%	4.2%
$ 4,000-5,700	1.2%	5.3%	5.3%	0%	5.7%	5.1%
$5,700-7,900	3.9%	6.6%	9.0%	0%	8.1%	8.6%
$7,900-10,400	7.2%	9.3%	13.3%	0%	12.6%	13.1%
$10,400-12,500	10.1%	11.0%	16.2%	0%	15.3%	16.2%
$12,500-17,500	26.3%	24.0%	30.3%	0%	30.2%	29.9%
$17,500-22,600	14.7%	13.3%	12.6%	0%	14.5%	13.5%
$22,600-35,500	10.3%	8.2%	3.7%	11.9%	5.0%	6.4%
$35,500-92,000	10.5%	7.5%	2.9%	34.6%	1.9%	2.0%
$92,000+	15.0%	10.7%	2.4%	53.5%	0.9%	1.0%
TOTAL	100.0%	100.0%	100.0%	100.0%	100.0%	100.0%

Exhibit III-28

Distribution of State and Local Taxes by Income Group:

Musgrave

Income Category	Personal Income Tax percentage of tax paid by Each Group; Base = $8.1b	Corporate Income Tax percentage of tax paid by Each Group; Base = $3.0b	General Sales Tax percentage of tax paid by Each Group; Base = $14.0b	Excise Tax percentage of tax paid by Each Group; Base = $8.7b	Gift & Estate Tax percentage of tax paid by Each Group; Base = $0.9b	Motor Vehicle Tax percentage of tax paid by Each Group; Base = $2.5b	Property Tax percentage of tax paid by Each Group; Base = $29.9b	Social Insurance Tax percentage of tax paid by Each Group; Base = $6.5b	TOTAL percentage of tax paid by Each Group; Base = $79.3b
$0-4,000	0%	4.1%	5.7%	4.9%	0%	4.8%	6.9%	1.0%	5.2%
$4,000-5,700	0.5%	5.3%	5.7%	6.0%	0%	5.2%	6.3%	2.6%	5.3%
$5,700-7,900	2.0%	6.6%	8.1%	9.6%	0%	9.8%	7.9%	6.2%	7.5%
$7,900-10,400	5.3%	9.3%	12.6%	13.9%	0%	12.1%	11.1%	12.0%	11.3%
$10,400-12,500	8.6%	11.0%	15.3%	16.6%	0%	15.3%	13.1%	15.2%	13.7%
$12,500-17,500	26.5%	24.0%	30.2%	29.2%	0%	31.1%	24.7%	30.9%	26.7%
$17,500-22,600	18.1%	13.3%	14.6%	11.9%	0%	12.4%	11.7%	18.0%	12.9%
$22,600-35,500	18.0%	8.2%	5.0%	3.5%	11.9%	4.8%	6.2%	11.5%	7.1%
$35,500-92,000	11.0%	7.5%	1.9%	2.4%	34.6%	2.1%	5.2%	1.7%	4.8%
$92,000+	10.0%	10.7%	0.9%	2.0%	53.5%	2.4%	6.9%	0.9%	5.5%
TOTAL	100.0%	100.0%	100.0%	100.0%	100.0%	100.00%	100.0%	100.0%	100.0%

92

group, while the burden of volume based charges may be distributed in line with water usage by income group. Water usage in turn may be calculated through estimates of the income elasticity of demand for water. This has been estimated in a number of studies, particularly Howe and Linaweaver, 1966.

The remaining task is twofold: first, determining the proportion of user charge revenues collected from industrial and residential users, and second, determining the composition of sewer charges between flat and volume based charges. With respect to the proportions between industrial/residential contributions of sewer charge revenues, some estimates may be found in the APWA report (for selected states). A weighted average of these yields 17% industrial and 83% residential sources.

Our survey results were similar; they indicate that approximately 20% of charge revenues is collected from industrial and commercial users. The remaining 80% is levied on residential users. The sewer bill is composed of 75% flat charges and 25% volume based charges on the average, according to an Urban Data Service publication, Sewer Services and Charges. The incidence calculations for the residential proportion of sewer charges are presented in Exhibit III-29.

This implies that for every $100 collected in sewer charges, $20 is passed along to consumers in the form of higher prices, and the remaining $80 is distributed between the ten income groups according to Column V of Exhibit III-29. The distribution of this burden is only slightly less regressive than a per capita or per household tax, because 75% of the charges are in the form of flat fees. The distribution of the property tax burden was included in the table for comparative purposes; the property tax is much more progressive than user charges. The distribution of the user charge and property tax burdens is depicted in Exhibit III-30. Reliance on user charges for financing water pollution abatement expenditures may be justified on the grounds of efficiency. Both the equity and efficiency properties of user charges should be evaluated in recommending financing mechanisms at the local level.

Incidence of Expenditure Substitution

If water pollution abatement costs are to be financed by reduction in other expenditures, then the distribution of the burden is equal to the distribution of benefits which would have accrued from the foregone expenditures. If, for example, expenditures on education are reduced to support the Act, then the burden of the Act falls on those people who would otherwise have benefited from education.

The Musgrave study, as well as the other work on expenditure incidence, begins by assuming that the dollar value of the gross benefits which result from government expenditures is equal to the

93

			Exhibit III-29		
	The Distribution of the Burden of Sewer Charges Levied On Residential Users				
I	II	III	IV	V	VI
Income Group	Water Consumption/ Family (gals/mo.)[1]	Distribution of Families[3]	Distribution of Water Consumption[4]	Distribution of Charge Burdens[5]	Distribution of Property Tax Burden[6]
1	3229	12.3%	6.6%	10.9%	6.9%
2	4221	7.0%	4.9%	6.5%	6.3%
3	4768	9.8%	7.7%	9.2%	7.9%
4	5493	12.0%	10.9%	11.7%	11.1%
5	5754	14.3%	13.6%	14.1%	13.1%
6	6421	24.5%	26.0%	25.0%	24.7%
7	7628	7.8%	9.8%	8.4%	11.7%
8	8597	8.5%	12.1%	9.4%	6.2%
9	12705	3.0%	6.3%	3.8%	5.2%
10	15000[2]	0.8%	2.0%	1.1%	6.8%
Total		100%	100%	100%	100%

[1]Water consumption/family was computed assuming an income elasticity of 0.4 (USR&E, "The Relationship Between Housing and Water Resources Planning and Management," for OWRR, March 1971). Average water consumption was assumed to be 6000 gallons/month (Urban Data Service, "Sewer Services and Charges," ICMA, February 1970, Vol. 2, No. 2).

[2]An arbitrary ceiling of 15,000 gallons/month was assumed.

[3]The distribution of families by income group was taken from Exhibit II-5.

[4]The distribution of water consumption was calculated from the average water consumption for each income bracket weighted by the number of families in each bracket.

[5]The distribution of the residential charge burden was calculated by assuming that 75% of residential sewer charges come from a flat charge, and 25% come from volume based charges (UDS, op.cit., Table 21 and Table 22.).

[6]The distribution of the property tax burden (included for comparative purposes) were taken from K.A. Musgrave, Karl E. Case and H. Leonard, "The Distribution of Fiscal Burdens and Benefits," HIER Discussion Paper #319, September 1973, Cambridge, Massachusetts.

Exhibit III-30

The Distribution of the Burden of Sewer Charges

Levied on Residential Users (Graph)

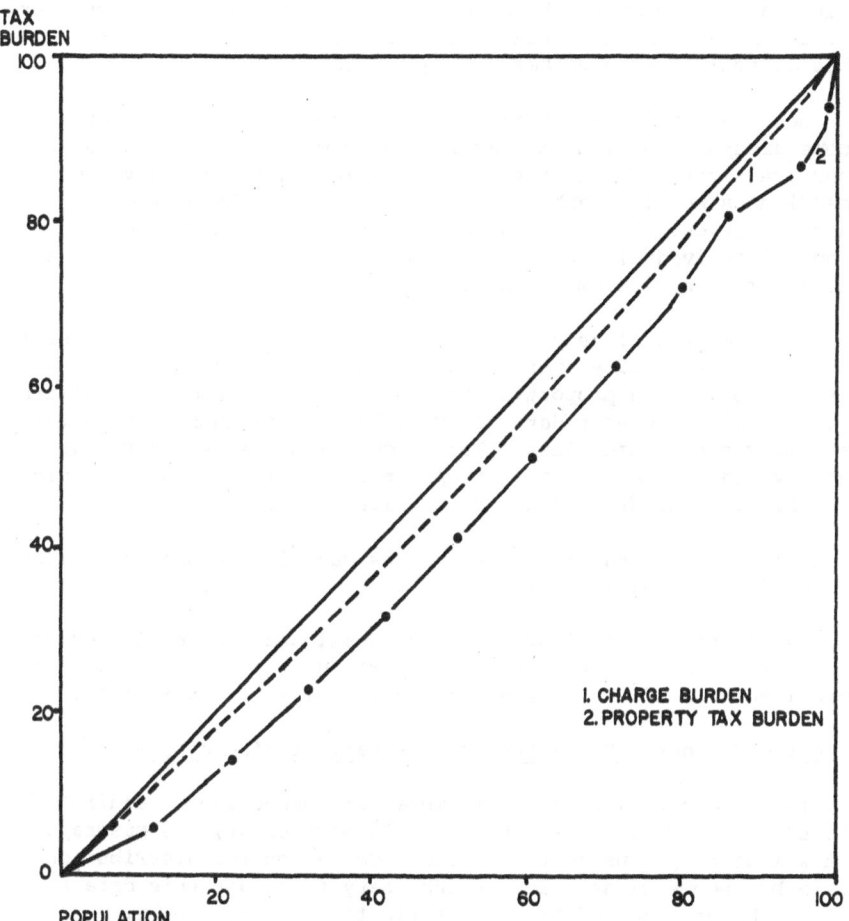

actual costs of those expenditures. This assumption simplifies the analysis considerably, since it allows Musgrave to concentrate on identifying the distribution of expenditures rather than the more amorphous distribution of realized benefits.

The next step in this study is to define and calibrate income groups, and to identify the absolute amount of total government expenditures made on particular programs. The expenditure data is taken directly from government budget data.

Next, expenditure figures are allocated among the selected income groups. As with the tax incidence work, a series of assumptions are first made about the way in which particular government expenditures benefit members of the population with particular characteristics. The mapping between these characteristics and the income of individuals is then determined, and used to distribute expenditure benefits among income groups.

An example will help to clarify this process. Musgrave assumes that the benefits of elementary and secondary school education accrue to families in proportion to the number of school-age children in that family. Census data is available on the number of children per family per income class. Total educational expenditures are thus divided up among income groups on the basis of the percentage of total U.S. children in any particular group.

A list of the shifting assumptions made in the Musgrave study is provided in Exhibit III-31.

Using this methodology, Musgrave developed some estimates of the distribution of the benefits of particular public expenditures among income groups. These estimates are given in Exhibit III-32.

Using the Incidence Estimates for the Various Financing Methods

The burden or incidence estimates discussed above require some adjustment before they can be used. In particular, the Musgrave data was developed using the national series on the distribution of people by income class. It was necessary to adjust this data to reflect the variance of income distributions across space.

A simple example will help clarify the problem. Suppose there are two income classes nationally, one with $5,000 and one with $45,000; and that each group is the same size. Suppose further that we are investigating the incidence of a tax which varies strictly with income. In this case, the low income group, with 10% of total income, pays 10% of the tax; the high income group bears the burden of the remaining 90%. Now suppose we wish to investigate the burden of an income tax imposed by a particular city. In this city, let us assume, only low income individuals live. Clearly, it would not be appropriate to apply national burden estimates to the city — the low

96

Exhibit III-31

Bases for the Allocation of Benefits of Government
Expenditures By Income Group

	ITEM	MUSGRAVE
1	National Defense	total income, total taxes, population
2	Postal Service	same as above
3	Health & Hospital	number of families; short stay hospital days
4	Police, Fire, Sanitation	same as Defense
5	Transportation & Commerce	same as Defense
6	General Government	same as Defense
7	Natural Resources, Recreation	same as Defense
8	Labor	same as Defense
9	Interest on Debt	interest income, dividend income
10	Housing	same as Defense
11	Veterans	veterans compensation & pension
12	Agriculture	farm income
13	Highways	auto expenditures, total consumption
14	Public Welfare	public assistance & social security
15	Education	elementary & secondary school enrollment, college

Exhibit III-32

Distribution of Government Expenditures by Income Category: Musgrave

PERCENTAGE OF VARIOUS EXPENDITURES RECEIVED BY EACH GROUP:

Income Category	Education	Higher Education	Elementary & Secondary Ed.	Highways	Health & Hospitals	Agriculture	Public Assistance	Social Insurance	Unemployment Compensation	Veterans	Interest	TOTAL
$0-4,000	4.2%	1.5%	5.1%	3.9%	20.2%	0%	79.7%	49.7%	12.0%	39.2%	5.5%	33.6%
$4,000-5,700	8.1%	5.5%	9.0%	5.7%	21.0%	1.5%	9.3%	14.7%	14.4%	12.0%	5.6%	11.1%
$5,700-7,900	12.9%	9.5%	14.0%	9.6%	16.7%	3.1%	5.5%	9.6%	15.1%	9.3%	5.1%	8.7%
$7,900-10,400	16.7%	15.0%	17.3%	13.2%	13.2%	5.7%	2.7%	6.6%	16.2%	9.7%	3.9%	7.6%
$10,400-12,500	16.4%	15.3%	16.7%	16.7%	9.6%	6.0%	2.0%	5.0%	13.6%	8.1%	5.0%	6.8%
$12,500-17,500	26.6%	29.4%	25.6%	29.7%	11.4%	13.2%	0.8%	8.6%	19.9%	13.6%	13.4%	12.2%
$17,500-22,600	8.3%	13.0%	6.7%	13.5%	5.3%	15.0%	0%	3.6%	5.8%	5.2%	8.9%	5.9%
$22,600-35,500	4.1%	6.5%	3.4%	3.9%	1.8%	20.4%	0%	1.6%	2.0%	2.1%	9.4%	4.3%
$35,500-92,000	2.0%	3.2%	1.6%	2.3%	0.9%	25.6%	0%	0.3%	1.0%	0.4%	10.5%	3.9%
$92,000+	0.7%	1.1%	0.6%	1.3%	0%	9.5%	0%	0.3%	0%	0.4%	32.8%	6.0%
TOTAL	100.0%	100.0%	100.0%	100.0%	100.0%	100.0%	100.0%	100.0%	100.0%	100.0%	100.0%	100.0%

income group will pay not 10% (as in the national case) but <u>100%</u> of the tax. Some account must be taken of the difference in the national and local income distributions.

Fortunately, a simple technique for making this adjustment was available* for use in this project. In particular, the share of benefits (or costs) in state/city i going to that state/city's income class k can be expressed as:

$$\frac{\beta ik}{\underset{k}{\Sigma}\beta ik} = \frac{(Pik/\underset{i}{\Sigma}Pik) \; [\underset{i}{\Sigma}\beta ik/\underset{ik}{\Sigma\Sigma}\beta ik]}{\underset{k}{\Sigma} \{ \; (Pik/\underset{i}{\Sigma}Pik) \; [\underset{i}{\Sigma}\beta ik/\underset{ik}{\Sigma\Sigma}\beta ik]}$$

where

$\dfrac{\beta ik}{\underset{k}{\Sigma}\beta ik}$ = share of benefits of state i going to income class k.

Pik = number of members in class k in state i

term in brackets = national share of program benefits going to income class k (Musgrave's estimates)

In order to apply this technique, we required the national incidence estimates and city/state data on income distributions provided in the Census. The result of this process was an estimate <u>for each financing tool</u>, and for each geographical subdivision, of the share of costs absorbed by each income group.

SURVEY OF LOCAL GOVERNMENTS FINANCING WATER POLLUTION CONTROL EXPENDITURES

In order to calculate the incidence of the costs of meeting the municipal treatment requirements of the Act, it is necessary to predict the method governments will select for financing their expenditures. One of the techniques involves analyzing the patterns and trends of federal, state and local government financing; this has been described earlier in this chapter.

Extrapolation of these trends could be misleading in light of the changes in municipal financing and budgeting induced by the federal water pollution control program. Therefore, as a check

---

*Charles Brown & James Medoff, "Revenue Sharing: The Share of the Poor," <u>Public Policy</u>, Spring, 1974.

on the fiscal trends derived from the Census of Government, we have conducted a survey of the municipalities, counties and districts which have constructed waterwaste treatment facilities.

The objective of the survey was to gather information on the methods of financing local governments use for their water pollution control expenditures. The results obtained are analyzed in order to test a number of hypotheses about possible factors influencing the method of financing selected, such as the size of the local share, the size of the local share relative to population size, type of government, type of project, and so on. The patterns estimated from the survey results are then compared with the patterns obtained from the analysis of local governments' fiscal patterns and trends.

The Grant Population

Under P.L.'92-500, local governments face mandatory treatment requirements and receive federal grant assistance for meeting these requirements. As P.L. 92-500 is almost three years old, there is some history of construction grants and projects under the Act. We have, therefore, included grants awarded under P.L. 92-500 in our grant population and sample.

Although the requirements of P.L. 92-500 are significantly different from those of previous legislation (P.L. 84-660), there are some strong similarities. In both cases, some treatment was required, and some federal assistance was available for the construction of facilities. Therefore, the experience of P.L. 84-660 is relevant for our investigation, and grants under that Act were included both in the grant population and sample.

In order to file for grant applications, local governments must formulate plans for financing the local share. Their plans are also relevant for our study; thus, we have included grants pending in our grant population and sample.

The grant population was limited to projects with total cost of at least $1 million. It was felt that with federal (and state) participation reducing the local share, the required total expenditures for lesser projects would be too small to be of relevance for our study. Thus, the grant population examined includes:

- under P.L. 84-660, all grants for projects (with minimum total cost of $1 million), awarded between January 1, 1967 and December 31, 1972. We felt this six year time period was sufficient for our purposes, and further, the appropriation of further monies in 1966 for treatment facility construction assistance served as a convenient cutoff point.

100

- under P.L. 92-500, all projects (with minimum total project cost of $1 million), awarded between January 1, 1973 and December 31, 1974.

- grants pending under P.L. 92-500, all grants (with minimum total project costs of $1 million), awaiting funding as of March 15, 1975.

Note that the construction projects described here do not include projects without federal assistance, as it seems unlikely that, under the current legislation, local governments will undertake construction projects without federal funding. This should not alter the significance of the results presented here.

The characteristics of the grant populations conform pretty much to expectations. The average project size was the smallest under P.L. 84-660, approximately $5.5 million, with the mean federal share amounting to 44% of this. Both the awarded and pending grants projects under P.L. 92-500 had substantially larger average total project costs, with the federal share approximating 75%.

In each category (92-500, 92-500 future, 84-600), grantees in the two smallest population categories (under 10,000 and 10-50,000) received most of the grants, with the local shares (including state contributions) clustered in the $.25 to $5 million range. Somewhat surprisingly, the size of the local share (which may be taken as an indicator of total project cost) did not show a strong association with population size. Analysis of the distribution of grants by the size of the local share and type of government indicates that in each case, municipalities (towns, cities, villages, and townships) received the most grants, indicating that they were the most active participants in the construction program at the local level. Special districts ranked second, with counties being the least active. No strong relationship between the type of government and the size of the local share was apparent. Finally, the distribution of grants by population size and type of government was analyzed. Municipal governments with small populations received the majority of the grants. This simply reflects the large number of small towns. In the case of the grants under P.L. 84-660, the tendency is reinforced by the provision of the law which earmarks at least 50% of the appropriated funds for smaller communities.

During the analysis of the grant population, we found that the grant population under P.L. 84-660 and P.L. 92-500 accounted for roughly 75-80% of the federal expenditures on the construction of sewerage facilities during the 1967-1974 period. In other words, while the $1 million lower limit for total project cost eliminated a large number of projects, only 20-25% of the federal program is excluded.

101

Sample Design

As described above, the total number in the grant population (with total project costs greater than $1 million) in various categories was as follows:

(1) P.L. 92-500 (awarded):   498 grants;

(2) P.L. 92-500 (future):   121 grants;

(3) P.L. 84-660 (1967-72): 1651 grants.

The first decision concerned the sampling frequency for each category of grants: 1, 2, and 3. We decided to send out 250 questionnaires in the first category, 60 in the second, and 300 in the third; this represents sampling frequencies of 1:2, 1:2, and 1:5.5, respectively. The choice of the greater sampling frequencies for categories 1 and 2 was based on a number of factors:

(1) Our task is to investigate the incidence of P.L. 92-500; therefore, the fiscal experience under that Act is the most relevant for our purposes.

(2) The amount of federal expenditures and fiscal obligations are far greater under P. L. 92-500 than under the previous Act; for the first two years of P.L. 92-500, 1973 and 1974, these figures were almost double the amounts for 1967-1972.*

The next decision concerned the sample selection for each category. We based our sample selection on the number of grants received, as this is representative of the breadth of fiscal experience; the amount of the local share was considered as a secondary factor. Thus, the number of questionnaires in each grant category was allocated between states proportional to the number of grants received by the states in each category. This may be expressed as

$$\frac{n_i}{N_i} = \frac{n}{N}$$

where   n = sample size

         N = population size

         i = sample stratum or state.

For example, in Category 1, New York State was awarded 28 grants; we have sent 14 questionnaires to examine their experience. We decided to stratify our sample on a state by state basis, rather than

_____

*U.S. Environmental Protection Agency, "Review of the Municipal Waste Water Treatment Works Program," November 1974, Washington, D.C.

by region, because there is little reason to believe that fiscal behavior is uniform within regions, while significant differences are likely to exist between states. The proportionality

$$\frac{n_i}{N_i} = \frac{n}{N}$$

was not observed rigorously. In order to increase coverage, the sampling frequency was increased for states which received relatively small numbers of grants, at the expense of states with large numbers.

The within-state allocation of questionnaires between population centers of different sizes was also decided on the basis of the number of grants received.

The proportionality in this case may be expressed as

$$\frac{n_{ik}}{N_{ik}} = \frac{n_i}{N_i}$$

where   n, N, I are as before

k = population size code.

If 20% of the grants awarded to a given state were received by centers with populations below 10,000 (population code #7), then approximately 20% of the state sample questionnaires were addressed to such centers. Again, the sampling frequency was increased for population categories which received a relatively small number of grants. Within a given population category, questionnaires were addressed to agencies on the basis of the size of the local shares. Although we attempted to cover the range of local share sizes, greater emphasis was given to larger sizes, particularly those received by the smaller centers.

In sum, our sample may be described as stratified and proportional. However, because the sample was designed systematically, it is not expected to be representative of the population. The increased sampling frequency for states and population centers which received a relatively small number of grants, and the selection of grants with large local shares, particularly when awarded to small population centers, is expected to introduce some bias into the sample.

Comparison of Sample and Grant Population Characteristics

Some of the grant population and sample characteristics have been summarized in Exhibits III-33 through III-36. Exhibits III-33 and III-34 describe the population and sample frequencies for two

Exhibit III-33

Distribution of Grants Between 1967 and 1974, by Population
Size and Type of Government

| Population Size (000s) | Type of Government | | | | Total |
	Municipalities	Counties	Special Districts	Not Known	
(1) 1000+	33	5	26	0	64
(2) 500-999	49	14	26	2	91
(3) 250-499	46	17	19	3	85
(4) 125-249	82	19	25	3	129
(5) 50-124	177	28	67	4	276
(6) 10-49	575	73	128	17	793
(7) <10	576	71	131	25	803
Size Not Known	3	5	16	5	29
Total	1541	232	438	59	2270

Exhibit III-34

Distribution of Grants Between 1967-1974, by Population
Size and Type of Government

(SAMPLE)

Population Size (000s)	Type of Government				Total
	Municipalities	Counties	Special Districts	Not Known	
(1) 1000+	8	2	10	0	20
(2) 500-999	11	6	10	1	28
(3) 250-499	11	10	8	1	30
(4) 125-249	27	6	10	2	45
(5) 50-124	51	7	22	1	81
(6) 10-49	133	17	29	7	186
(7) <10	148	18	39	6	211
Size Not Known	1	-	6	1	8
Total	390	66	134	19	609

independent variables: population size and type of government. According to these tables, most of the grants were received by communities in the two smallest population categories: under 10,000 and 10,000-50,000. The grant population and sample proportions were compared in order to evaluate the "representativeness" of the sample. The chi-squared statistics calculated on the basis of these tables were 20.64, with 20 degrees of freedom. This is insufficient to reject the null hypothesis that sample proportions are similar to the proportions of the grant population at any reasonable level of confidence. In other words, the sampling bias described above was not statistically significant.

Exhibit III-35 describes the distribution of grants by the size of the local share, for both the grant population and the sample. The local share of the majority of grants was between $.5 and $5 million. In the sample, these categories accounted for 67.2% of all grants; in the population, this figure was 68.7%.

In the samples, municipalities received more grants than special districts and counties. Project costs, federal funding, and local shares are compared for the grant population and sample in Exhibit III-36. This table reflects the sampling bias: the project costs, the magnitude of federal funding and of the local shares are greater in the samples than in the grant population.

Survey Results

The response rate on the questionnaire was 37%; some 226 completed questionnaires were received. In order to find out whether the response was biased, respondent and sample proportions were compared in Exhibits III-37 and III-38. By population size, the proportions of the sample and respondents were roughly similar. The chi-squared statistic was 3.076 with 8 degrees of freedom, indicating that the null hypothesis of the respondent frequencies were similar to the sample frequencies and could not be rejected at any of the normal significance levels. Sample and respondent proportions by type of place, however, were dissimilar; the frequency of response was higher for special districts than expected, and lower for municipalities and counties. These differences were statistically significant, the chi-squared statistic was calculated at 16.48 with 2 degrees of freedom.

The survey instrument, the questionnaire, was divided into four major parts. In the first part, the information was gathered on the capital costs associated with the project, and on the federal and state shares. In addition, the initial method of payment was considered in this section. Questions were asked about the types of bonds used, their relative importance, interest rates, and maturity dates. The second part addressed the question of ultimate repayment; information was gathered on the types of revenue sources

Exhibit III-35

Percentage Distribution of Grants by Size of Local Share,

Grant Population and Sample

Size of Local Share ($Million)	Grant Population	Sample
(1) $0-.25	1.5%	1.5%
(2) $.25-.5	.5%	9.6%
(3) $.5-1	12.9%	22.2%
(4) $1-2	29.8%	22.9%
(5) $2-5	26.0%	22.1%
(6) $5-10	18.3%	10.4%
(7) $10-25	5.7%	9.2%
(8) $25+	4.0%	2.1%
TOTAL	100.0%	100.0%

Exhibit III-36

Comparison of Costs for

Grant Population and Sample

(costs in $000)

	92-500 Sample	84-660 Sample	92-500 Future Sample
Mean Project Cost	$10,098	$9,065	$8,544
Mean Federal Funding	$ 7,241	$3,491	$6,213
Mean Local Share	$ 2,833	$5,594	$2,342
	92-500 Grant Population	84-660 Grant Population	92-500 Future Grant Population
Mean Project	$ 8,327	$5,581	$7,685
Mean Federal Funding	$ 6,002	$2,431	$5,824
Mean Local Share	$ 2,237	$3,258	$2,039

Exhibit III-37

Sample and Respondent Frequency
by Population Size

Population Size (000s)	Responses	% Dist.	Total Sample	% Dist.
(1) 1000+	5	2.2%	20	3.3%
(2) 500-999	11	4.9%	28	4.6%
(3) 250-499	15	6.6%	30	4.9%
(4) 125-249	18	8.0%	45	7.4%
(5) 50-124	32	14.1%	81	13.3%
(6) 10-49	72	31.9%	186	30.5%
(7) <10	71	31.4%	211	34.6%
Size Not Known	2	0.9%	8	1.3%
Total	226	100%	609	100%

Exhibit III-38

Sample and Respondent Frequency
by Type of Government

Type of Government	No. of Responses	% Dist.	Total Sample	% Dist.
Special District	76	33.6%	134	22.0%
Municipality	133	58.8%	390	64.0%
County	17	7.5%	66	10.8%
Not Known	--	--	19	3.1%
Total	226	100%	609	100%

used by local governments for meeting their annual capital cost
obligations. The revenue instruments considered included the
usual sources of tax revenues, with expenditure reductions also
being considered as a potential revenue source. As the method of
financing was expected to differ between the capital and O&M portions of the costs, in the third section similar questions were
addressed for O&M costs. In the final section of the questionnaire, information was gathered on legal issues such as debt
ceilings, interest rate ceilings, and taxing powers.

The mean capital cost of the projects of the respondents
was $11.59 million. This may be compared with the mean project
costs for the sample: $9.43 million. This indicates a further
bias in the response; with larger projects being more heavily
represented among the respondents. The relationship of project
cost, population size and type of government is discussed in
Exhibit III-39 and III-40 respectively. Overall, total project
cost does increase with population size, from $5.1 million for
the smallest centers, to $48.8 million for the largest. Costs for
the medium sized cities (with populations of 250,000-1,000,000)
decreased, according to the data presented here. One explanation
for this phenomenon is that the middle sized cities have already
built most of their facilities. Therefore, the projects undertaken
at the present time and in the recent past are more limited in
scope. Alternatively, the cost decrease could have been caused by
an anomaly in the data for these cities. According to Exhibit
III-40, the average project undertaken by municipalities was the
most expensive, with special districts and counties undertaking
less costly projects.

Exhibit III-39	
Sample's Project Cost (Total Capital) by Population Size	
Population Size (000s)	Mean Project Cost ($000s)
(1) 1000+	$48,757
(2) 500-999	11,777
(3) 250-499	19,962
(4) 125-249	26,656
(5) 50-124	12,938
(6) 10-49	9,433
(7) <10	5,156
AVERAGE COST	$11,587

Exhibit III-40

Sample's Mean Project Cost (Total Capital)

by Type of Government

Type of Government	Mean Project Cost ($000s)
Special District	$11,703
Municipality	12,350
County	5,099
AVERAGE COST	$11,587

The federal, state and, therefore, local shares may be expected to vary between the three different types of government. Exhibit III-41 presents these estimates. The average federal share for the respondents was 66.2%, the average state and local shares were 13.3% and 20.5% respectively.

According to the information in Exhibit III-41, projects undertaken by counties had the highest federal and lowest state shares. The high federal share for counties may simply indicate that most of their projects received grant assistance under P.L. 92-500.

Exhibit III-41

Federal, State and Local Shares in Sample

by Type of Government

Type of Government Government	Federal Share	State Share	Local Share
Special District	50.2%	21.5%	28.5%
Municipality	70.9%	8.6%	20.5%
County	73.6%	6.8%	19.6%
AVERAGE SHARE	66.2%	13.3%	20.5%

This finding, and other findings pertinent to counties, should be accepted with some caution, as they are based on 17 responses from counties. Municipalities received the second highest level of federal funding, and a below-average rate of state assistance. The opposite is true for special districts. One explanation for this phenomenon is that special districts undertake projects which include expenditures that are not federally funded, such as collector construction and land acquisition.

Annual O&M costs represented approximately 3.76% of the capital

cost of the facility. The data did not show any significant relationship between the ratio of O&M and capital costs and the size
of place or type of government.

With respect to debt financing, the average interest rate reported was 5.39%, and the average length of maturity for bonds was
23.28 years. Again, no significant relationships with size of
place or type of government were apparent from the data.

The use of the various debt instruments — general obligation
bonds, special assessment bonds, and so on — is described in Exhibits III-42 and III-43.

The most frequently used debt instrument was the general obligation bond, with almost half the respondents reporting its use.
Other revenues — primarily accumulated reserves — were used by
approximately a third of the respondents, and almost a third reported using sewer revenue bonds. The use of the other debt instruments
was negligible. The use of the various debt instruments did not
vary significantly with type of government or size of place.

The use of the debt instruments should be compared with the
authority to issue various forms of debt: the relevant information
is presented in Exhibit III-44. Of the 164 respondents who had the
authority to issue general obligation bonds, 67% or 110, used them.
With respect to sewer revenue bonds, 137 had the necessary authority, but only 43%, or 59 respondents, used them. Most municipalities and special districts had the authority to use general obligation bonds, and a large proportion of both could issue sewer revenue
bonds as well. County governments were largely restricted to the
use of sewer revenue bonds. The authority to issue various types of
debt did not vary systematically with the size of place.

Of the total value of the initial (debt) financing, general
obligation bonds accounted for 48%, sewer revenue bonds for 28%,
accumulated reserves for 17%, and miscellaneous debt for the remaining 7%.

The use of various taxes for the ultimate financing of the
local share of the capital cost (repayment of debt) is described
in Exhibits III-45 and III-46. The most frequently used revenue
source for financing the local annual capital cost was the sewer
charge: 122 respondents reported its use. By contrast, only 57
authorities relied on the property tax, 36 used water charges,
and 18 used special assessments. The use of sewer charges did not
vary between municipalities, special districts, and counties.
Approximately 37% of the special districts reported using the
property tax — in contrast to less than 24% of counties and muni-
ciplaties.

On the basis of the information from the Census of Government,
it might be anticipated that the structure of tax revenues will

Exhibit III-42

The Use of Debt Instruments

by Type of Government

Type of Government	General Obligation Bond		Special Assessment Bond		Sewer Revenue Bond		Other Revenue Bond		Other*	
	Used	Not Used	Used	Not Used	Used	Not Used	Used	Not Used	Used	Not Used
Special District	37	39	4	72	20	56	10	66	23	53
Municipality	65	68	6	127	34	99	9	124	41	92
County	8	9	0	17	5	12	0	17	7	10
All Respondents	110	116	10	216	59	167	19	207	71	155

*This category consists almost entirely of accumulated reserves.

Exhibit III-43

The Use of Debt Instruments

by Population Size

Population Size (000s)	General Obligation Bond		Special Assessment Bond		Sewer Revenue Bond		Other Revenue Bond		Other*	
	Used	Not Used	Used	Not Used	Used	Not Used	Used	Not Used	Used	Not Used
1000+	2	3	0	5	2	3	0	5	2	3
500-999	3	8	0	11	4	7	1	10	4	7
250-499	8	7	1	14	4	11	0	15	4	11
125-249	10	8	0	18	4	14	1	17	7	11
50-124	18	14	1	31	6	26	2	30	13	19
10-49	36	36	2	70	21	51	7	65	15	57
<10	33	38	6	65	17	54	8	63	25	46
Size Not Known	0	2	0	2	1	1	0	2	1	1
All Respondents	110	116	10	216	59	167	19	207	71	155

*This category consists almost entirely of accumulated reserves.

Exhibit III-44

Authority to Issue Debt Instruments by Type of Government

Type of Government	General Obligation Bonds		Special Assessment Bonds		Sewer Revenue Bond		Other Revenue Bond	
	Yes*	No**	Yes	No	Yes	No	Yes	No
Special District	54	20	21	53	44	30	17	57
Municipality	99	29	37	91	84	44	17	11
County	1	15	3	13	9	7	2	14
All Respondents	164	54	61	157	137	81	36	182

*Had authority to issue.

**Did not have authority to issue.

NOTE: These columns do not total to 226 because Section IV of the questionnaire was not completed for 8 questionnaires.

Exhibit III-45

Financing the Local Share of Capital Costs: The Method of Repayment by Type of Government

Type of Government	Property Tax		Special Assessment		General Sales Tax		Special Sales Tax		Local Income Tax		Industrial Contribution	
	Used	Not Used	Used	Not Used	Used	Not Used	Used	Not Used	Used	Not Used	Used	Not Used
Special District	28	48	10	66	0	76	0	76	0	76	1	75
Municipality	25	108	8	125	1	132	1	132	0	133	6	127
County	4	13	0	17	0	17	0	17	0	17	0	17
All Respondents	57	169	18	208	1	225	1	225	0	226	7	219

Type of Government	Sewer Charge		Water Charge		Other Charge		Contribution from other Towns		Other	
	Used	Not Used	Used	Not Used	Used	Not Used	Used	Not Used	Used	Not Used
Special District	40	36	10	66	5	71	1	75	3	73
Municipality	72	61	24	109	4	129	5	128	8	125
County	10	7	2	15	0	17	0	17	0	17
All Respondents	122	104	36	190	9	217	6	220	11	215

Exhibit III-46

Financing the Local Share of Capital Costs:
The Method of Repayment by Population Size

Population Size	Property Tax		Special Assessment		General Sales Tax		Special Sales Tax		Local Income Tax		Industrial Contribution	
	Used	Not Used	Used	Not Used	Used	Not Used	Used	Not Used	Used	Not Used	Used	Not Used
1000+	2	3	0	5	0	5	0	5	0	5	0	5
500-999	0	11	0	11	0	11	0	11	0	11	0	11
250-499	4	11	0	15	0	15	1	14	0	15	2	13
125-249	4	14	2	16	0	18	0	18	0	18	0	18
50-124	10	22	3	29	0	32	0	32	0	32	0	32
10-49	18	54	4	68	0	72	0	72	0	72	1	71
<10	18	53	9	62	1	70	0	71	0	71	4	67
Size Not Known	1	1	0	2	0	2	0	2	0	2	0	2
All Respondents	57	169	18	208	1	225	1	225	0	226	7	219

Population Size	Sewer Charge		Water Charge		Other Charge		Contribution from other Towns		Other	
	Used	Not Used	Used	Not Used	Used	Not Used	Used	Not Used	Used	Not Used
1000+	2	3	0	5	0	5	0	5	0	5
500-999	8	3	1	10	1	10	0	11	0	11
250-499	7	8	4	11	0	15	0	15	0	15
125-249	9	9	2	16	1	17	0	18	0	18
50-124	16	16	3	29	2	30	1	31	2	30
10-49	39	33	13	59	3	69	3	69	6	66
<10	39	32	12	59	2	69	2	69	3	68
Size Not Known	2	0	1	1	0	2	0	2	0	2
All Respondents	122	104	36	190	9	217	6	220	11	215

116

vary with size of place. In particular, the larger places are expected to demonstrate diversity, to rely on a larger number of revenue sources. This is not the case for financing the capital costs of sewerage facilities — cities in the three largest size categories rely almost exclusively on sewer charges and water charges. The smaller cities use these but also use property taxes and special assessments to a lesser extent.

The authority to levy taxes is described in Exhibits III-47 and III-48. Almost all governments with the authority to levy sewer charges (193) use them (190). Of the 148 local governments with the authority to levy property taxes, only 57 do so for this purpose.

As anticipated, municipalities have the broadest taxing powers. These, however, do not vary with city size; small and large centers have similar powers to levy taxes.

Sewer charges represent the most significant revenue source for financing local capital costs; they account for some 47%. Water charges are used to finance another 28%, property taxes account for 20%. The balance comes from miscellaneous sources.

The financing of operations and maintenance charges is described in Exhibits III-49 and III-50. According to these tables, the most widely used revenue source for financing O&M costs is the sewer charge: 166 respondents indicated their use. Industrial contributions were used almost as extensively (149), while only 42 reported relying on the property tax for this purpose. Water charges were used to a limited extent — 24 respondents indicated their use. General and special sales taxes and the local income tax were not used for this purpose at all. County governments relied almost exclusively on industrial contributions and sewer charges; special districts and municipalities both used the property tax as well. No systematic variation between the method of financing for O&M costs and city size was apparent from the data. Of the total volume of O&M costs, sewer charges financed 57%, property taxes 20%, and industrial contributions another 14%. The balance came from miscellaneous sources. These figures should be compared with the method of financing for local capital costs — where the property tax also accounted for 20%, sewer charges for 47%, and water supply charges for 28%. The methods of O&M and capital financing at the local level are remarkably similar.

Another interesting feature of the survey was the response concerning expenditure reductions for financing sewerage facilities. Almost 10% of the respondents indicated that they have or will undertake expenditure reductions in order to have adequate revenue. On the basis of this response, it was not possible to estimate the proportion of total costs financed from expenditure reductions. The types of services reduced, however, were indicated; these included other sewerage services, such as stormwater collection; other

Exhibit III-47

The Authority to Levy Taxes by Type of Government

Type of Government	Property Tax Yes*	Property Tax No**	Special Assessment Tax Yes	Special Assessment Tax No	General Sales Tax Yes	General Sales Tax No	Special Sales Tax Yes	Special Sales Tax No	Local Income Tax Yes	Local Income Tax No	Sewer Charges Yes	Sewer Charges No	Water Charges Yes	Water Charges No
Special District	51	24	45	30	13	62	4	71	6	69	66	9	48	27
Municipality	86	44	64	66	22	108	10	120	4	126	112	18	69	61
County	11	6	5	12	3	14	2	15	0	17	15	2	11	6
All Respondents	148	74	119	108	38	184	16	206	10	212	193	29	128	94

*Has authority to issue.

**Does not have authority to issue.

118

Exhibit III-48

The Authority to Levy Taxes by Size of Place

Population Size of Place (000)	Property Tax		Special Assessment Tax		General Sales Tax		Specific Sales Tax		Local Income Tax		Sewer Charges		Water Charges	
	Yes*	No**	Yes	No	Yes	No	Yes	No	Yes	No	Yes	No	Yes	No
1000+	3	2	1	4	0	5	0	5	1	4	5	0	1	4
500-999	8	3	6	5	3	8	2	9	0	11	11	0	7	4
250-499	10	4	6	8	2	12	3	11	0	14	11	3	7	7
125-249	9	9	11	7	3	15	2	16	2	16	13	5	10	8
50-124	20	12	16	16	4	28	3	29	2	30	27	5	13	19
10-49	51	19	40	30	12	58	2	68	1	69	59	11	42	28
<10	46	24	35	35	14	56	3	67	4	66	65	5	46	24
Size Not Known	1	1	0	2	0	2	1	1	0	2	2	0	2	0
All Respondents	148	74	119	107	38	184	16	206	10	212	193	29	128	94

*Has authority to issue.

**Does not have authority to issue.

119

Exhibit III-49

Financing O&M Charges by Type of Government

Type of Government	Property Tax Yes*	No*	Special Assessment Tax Yes	No	General Sales Tax Yes	No	Special Sales Tax Yes	No	Local Income Tax Yes	No	Industrial Contribution Yes	No	Sewer Charges Yes	No	Water Charges Yes	No	Other Charges Yes	No	Contributions from other towns Yes	No	Other Yes	No
Special District	19	57	0	76	0	76	0	76	0	76	47	29	51	25	7	69	7	69	5	71	8	68
Municipality	22	111	3	130	0	133	0	133	1	132	91	42	101	32	16	117	11	122	14	119	11	122
County	1	16	0	17	0	17	0	17	0	17	11	6	14	3	1	16	2	15	1	16	0	17
All Respondents	42	184	3	223	0	226	0	226	1	225	149	77	166	60	24	202	20	207	20	206	19	207

*Used **Did not use.

Exhibit III-50
Financing O&M Charges by Size of Place

Size of Place (000)	Property Tax Yes*	Property Tax No**	Special Assessment Tax Yes	Special Assessment Tax No	General Sales Tax Yes	General Sales Tax No	Specific Sales Tax Yes	Specific Sales Tax No	Local Income Tax Yes	Local Income Tax No	Industrial Contribution Yes	Industrial Contribution No	Sewer Charges Yes	Sewer Charges No	Water Charges Yes	Water Charges No	Other Charges Yes	Other Charges No	Contributions from other Towns Yes	Contributions from other Towns No	Other Yes	Other No
1,000+	3	2	0	5	0	5	0	5	0	5	5	0	3	2	0	5	0	5	0	5	1	4
500-999	2	9	0	11	0	11	0	11	0	11	8	3	7	4	2	9	1	10	1	10	0	11
250-499	3	12	0	15	0	15	0	15	0	15	10	5	11	4	2	13	1	14	2	13	0	15
125-249	3	15	1	17	0	18	0	18	0	18	13	5	15	4	2	16	2	16	0	18	1	17
50-124	8	24	2	30	0	32	0	32	0	32	20	12	21	11	3	29	4	28	2	30	3	29
10-49	15	57	0	72	0	72	0	72	0	72	42	30	51	21	7	65	7	65	8	64	8	64
<10	8	63	0	71	0	71	0	71	1	70	50	21	57	14	8	63	4	67	7	64	6	65
Size Not Known	0	2	0	2	0	2	0	2	0	2	1	1	2	0	0	2	1	1	0	2	0	2
All Respondents	42	184	3	223	0	226	0	226	1	225	149	77	166	60	24	202	20	206	20	206	19	207

*Used. **Did not use.

121

public works programs, road paving, parks and recreation services; and in one instance, the entire town budget, including fire and police protection and the construction of a new school.

INCIDENCE CALCULATIONS

Calculating the distribution of the expenditure on publicly owned wastewater treatment facilities involves a number of operations, as presented in Exhibit III-51. The operations are as follows:

1) Multiply the cost matrix I by the federal/state/local shares, matrix III. For local shares calculate annualized costs.

2) Multiply the baselines, matrix II by their federal/state/ local shares, matrix IV. For local shares calculate annualized costs.

3) Take the difference of these products to get the net costs by type of government.

4) Multiply the remainder by matrix V, which projects the method of financing, to get the distribution of costs by type of financing.

5) Multiply the resulting product by matrix VI, which gives the distribution of tax burdens/expenditure reductions by income group, to get the initial burdens by income group.

6) Adjust the "national" distribution pattern for each cell on the basis of its income distribution: matrix VII, with the adjustment being made through the Brown & Medoff formula.

7) The result is matrix VIII which is the distribution of burdens by income group.

Note that this set of operations is completed twice, once for capital costs, and once for O&M costs because the federal/state/local shares are clearly different for O&M and capital costs. The two sets of burdens (capital and O&M) are then summed to yield a single distribution pattern.

Each of the eight matrices are described below in terms of their form and alternative assumptions behind them.

It should be noted that the specification of the cost and baseline matrices (I and II) and the method of financing, matrix V, were based on two sets of data: 1) the information on federal, state and local government fiscal structures and trends contained in the section of this chapter dealing with this subject; and 2) the data obtained from the survey of local governments which

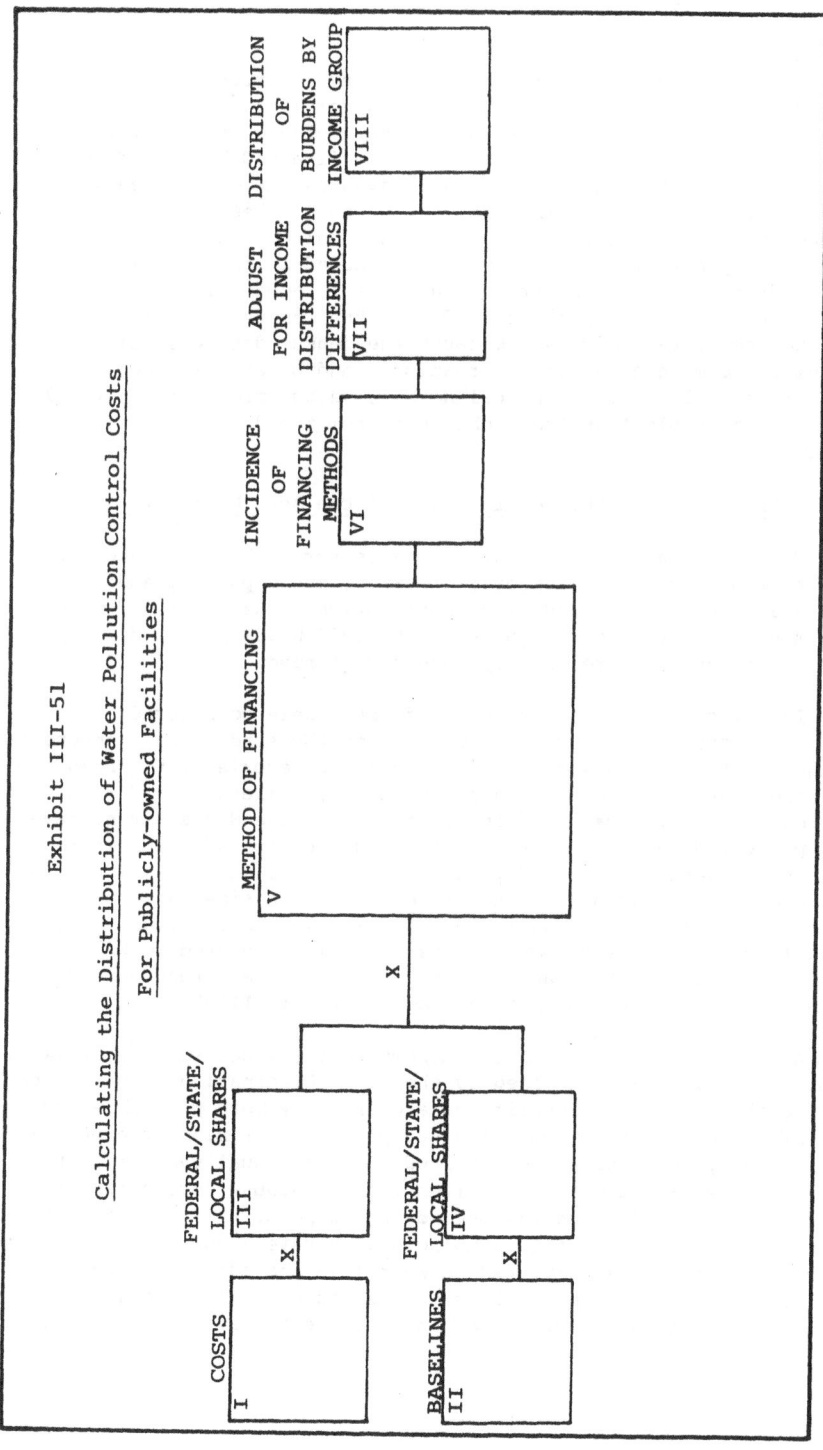

Exhibit III-51

Calculating the Distribution of Water Pollution Control Costs

For Publicly-owned Facilities

COSTS
I

FEDERAL/STATE/
LOCAL SHARES
III

BASELINES
II

FEDERAL/STATE/
LOCAL SHARES
IV

METHOD OF FINANCING
V

INCIDENCE
OF
FINANCING
METHODS
VI

ADJUST
FOR INCOME
DISTRIBUTION
DIFFERENCES
VII

DISTRIBUTION
OF
BURDENS BY
INCOME GROUP
VIII

have financed wastewater treatment facility construction under
various grants programs of the federal government, described above.
These data sources complement one another. The question of the extent to which the treatment costs will be financed by expenditure
reductions rather than by additional revenues can only be answered
using survey data. Likewise, for the parts of these costs financed
through expenditure reductions, the best source of data on which
items of expenditure are likely to be reduced is the information
gained from our questionnaires. In general, however, more weight
was given the information obtained from the Census of Governments
than to the survey results. Because the census data was based on a
presumably complete set of observations, and because it contained
both historical and geographical detail, we assumed that its results
were more reliable than those obtained from the 226 responses of the
survey.

The Costs of Meeting the Requirements of the Act (Matrix I)

Matrix I contains the cost estimates for each state and for
each type of government (special district, municipality, and county)
within each state. In addition, cost estimates are broken down by
five municipality size categories. The 1974 Needs Survey developed
cost estimates by state and by category of need.

In our calculations, two sets of costs were considered:
(1) categories I, II and IVB, composed of the traditional components
of the construction grant program, and (2) categories I-V, including
all needs categories except category VI, the collection and treatment
of stormwater. The capital cost estimates of the Needs Survey have
been presented in Exhibit III-2. The distribution of expenditures
between the various types of government (special district, municipality, and county) and between the five population size categories was
assumed to remain as it was in 1972. The 1972 distribution was calculated from the Census of Governments; this is presented in Exhibits
III-52 and III-53. The "needs" costs for each state then were distributed in line with the percentages of Exhibit III-53.

According to one scenario, the expenditures described above will
occur before 1985 as specified by the Act. Alternative scenarios concerning the timing of expenditures are probably more plausible; for
some of these, see the Commission's report. The timing of expenditures clearly influences the magnitude of the annual capital costs
which must be financed for any given year. Because forecasting the
time path of expenditures was beyond the scope of this effort, and
for reasons of computational simplicity, we have assumed that all construction activity is completed between 1973 and 1983, and that construction takes place in equal, discrete segments each year, i.e.,
that 1/10 of the total cost given in the Needs Survey is incurred each
year.

$$AE_{ij} = \frac{TC_{ij}}{10}$$

where AE_{ij} is the annual expenditure for a given type/size of government in a given state, and TC_{ij} is the total cost for a given type/size government in a given state.

Federal, State and Local Shares (Matrices III and IV)

To obtain the federal, state, and local shares of the invest-ments undertaken by each type of local government, matrix I was multiplied by matrix II, the federal, state, and local shares. For the purposes of this study, the federal share was assumed to be constant at 75%, state shares were assumed to remain at their current magnitudes, with the remainder being the local share. Estimates of the state shares are presented in Exhibit III-54. These estimates have been developed by Meta Systems, Inc. in a study for EPA entitled, The Evaluation of Methods of Financing Municipal Waste Treatment Works.

The local share of the annual capital expenditures is fi-nanced through a bond issue or from accumulated surplus. The relevant annual cost figure should, therefore, include interest cost.

The federal, state, and local shares vary with the nature of the costs: this is, capital costs under P.L. 92-500, expendi-tures that would have taken place without the Act, and operating and maintenance costs.

Annualized Capital Costs: Although a typical bond issue involves a mixture of bonds with differing maturities and interest rates, their repayment is normally channeled through a sinking fund. Therefore, the relevant annual cost figure may be computed through the sinking fund formula:

$$AC_{ij} = \frac{TC_{ij}}{10} \frac{r}{(1 + r)^n - 1} + r$$

where

AC_{ij} = annual cost for a given type of government in a given state,

n = maturity (we are assuming an average maturity of 20 years)

r = interest rate.

Although interest rates do vary over time, we have no fore-casts for this. Therefore, temporal variations in the interest rate are ignored. County, special district, and municipal govern-ments typically issue different types of debt — with different interest rates. Further, the credit rating of municipalities

125

Exhibit III-52

Total Sewerage Expenditures -- By Governmental Unit and State -- 1972 ($Millions)

State	Total	Counties	Municipalities (all Sizes)	Municipality Size Group (000s)					Special District
				100+	25-100	10-25	5-10	less than 5	
Alabama	$ 17.22	$ 5.06	$ 12.15	$ 5.69	$ 2.76	$ 1.66	$ 1.09	$ 0.95	$ 0
Alaska	16.95	12.67	4.27	0	0.22	2.57	0.31	1.06	0
Arizona	14.87	1.60	12.95	8.59	2.74	0.81	0.59	0.21	0.32
Arkansas	8.68	0	9.57	2.55	3.36	1.42	0.60	0.64	0.11
California	285.35	16.02	168.06	90.80	49.95	24.18	5.01	5.11	101.27
Colorado	41.17	0.09	27.73	13.24	7.59	4.89	1.23	0.78	13.34
Connecticut	87.46	0	72.99	18.27	25.37	19.13	8.42	1.00	15.17
Delaware	17.54	13.18	4.36	0	3.29	0.60	0.19	0.28	0
District of Columbia	36.32	0	36.32	36.32	0	0	0	0	0
Florida	76.86	11.50	64.67	29.21	18.35	9.61	3.46	4.05	0.69
Georgia	42.83	10.74	31.49	15.47	4.29	4.25	1.28	6.20	0
Hawaii	16.90	1.99	14.91	14.91	0	0	0	0	0
Idaho	2.72	0.05	2.42	0	1.00	0.60	0.21	0.61	0.25
Illinois	166.68	1.11	71.60	22.66	19.94	15.20	7.37	6.43	93.98
Indiana	83.68	0.09	83.26	22.60	31.33	17.69	5.14	6.49	0.34
Iowa	27.76	0	27.52	5.30	13.36	2.82	2.92	3.13	0.25
Kansas	15.19	0.01	15.17	8.79	1.56	2.65	1.12	1.06	0.01
Kentucky	25.07	2.08	21.27	12.40	1.16	2.24	3.70	1.76	1.72
Louisiana	28.04	2.61	25.16	16.10	4.78	1.61	1.92	0.76	0.27
Maine	8.14	0	6.57	0	0.87	2.08	2.71	0.91	1.57
Maryland	98.22	26.45	25.85	19.01	3.33	0.94	1.85	0.71	45.92
Massachusetts	89.90	0	89.39	13.42	32.93	32.41	5.35	5.27	0.51
Michigan	322.71	151.69	165.21	94.07	45.29	7.87	10.58	7.39	5.81
Minnesota	114.00	0.07	88.71	23.24	27.86	15.73	8.64	13.24	25.22
Mississippi	14.17	0.04	14.13	0.71	9.80	1.61	0.85	1.16	0

Exhibit III-52 (Cont'd)

Total Sewerage Expenditures -- By Governmental Unit and State -- 1972 ($Millions)

State	Total	Counties	Municipalities (all Sizes)	Municipality Size Group (000s)					Special District
				100+	25-100	10-25	5-10	less than 5	
Missouri	$ 41.95	$ 0.02	$ 27.57	$ 13.50	$ 3.90	$ 4.45	$ 2.53	$ 3.19	$ 14.36
Montana	3.09	0.67	2.39	0	1.41	0.33	0.12	0.53	0.03
Nebraska	21.15	0	19.01	11.87	1.92	3.00	0.90	1.33	2.14
Nevada	6.77	0.05	4.14	0.90	1.74	1.24	0.14	0.11	2.57
New Hampshire	4.88	0	4.86	0	2.42	0.94	1.01	0.49	0.01
New Jersey	180.05	1.69	86.44	12.38	20.80	18.86	25.56	8.85	91.92
New Mexico	6.46	0.17	6.29	4.36	0.87	0.51	0.29	0.26	0
New York	461.08	153.01	308.07	178.75	56.63	41.31	21.33	10.05	0
North Carolina	39.24	4.81	33.25	12.68	11.03	4.88	2.33	2.33	1.18
North Dakota	4.17	0	4.16	0	3.72	0.20	0.16	0.09	0
Ohio	175.25	30.82	143.50	66.98	38.98	22.25	9.35	5.94	0.93
Oklahoma	13.42	0	13.16	6.86	2.14	1.74	0.87	1.56	0.26
Oregon	29.38	2.14	24.45	9.04	4.95	4.92	3.21	2.34	2.79
Pennsylvania	207.56	1.75	74.63	27.28	15.65	12.72	11.44	7.04	131.19
Rhode Island	14.38	0	8.41	1.46	4.85	2.00	0.07	0.04	5.97
South Carolina	36.87	1.01	17.03	8.39	3.03	1.65	3.09	0.87	8.82
South Dakota	2.14	0	2.23	0	1.26	0.34	0.43	0.21	0.17
Tennessee	47.78	0.06	46.67	32.14	2.01	5.90	3.04	3.57	1.05
Texas	113.45	0	96.59	56.69	18.44	11.41	4.61	5.43	16.87
Utah	7.16	0.25	3.67	0.60	1.00	0.73	0.82	0.52	3.23
Vermont	5.99	0	5.99	0	0.91	1.05	2.75	1.28	0
Virginia	69.53	25.96	35.18	23.69	4.16	3.74	2.20	1.40	8.39
Washington	69.36	0.21	47.21	32.22	3.80	7.26	1.84	2.09	21.95
West Virginia	10.34	0	8.76	0	4.27	1.93	0.90	1.66	1.57
Wisconsin	114.87	0.04	75.57	18.44	22.03	22.09	7.10	5.90	39.26
Wyoming	1.21	0	1.10	0	0.64	0.16	0.15	0.14	0.11

Exhibit III-53

Total Sewerage Expenditures -- 1972 -- % Distribution by Governmental Unit and State

State	Total	Counties	Municipalities (all sizes)	Municipality Size Group (000s)					Special District
				100+	25-100	10-25	5-10	<5	
Alabama	100%	29.40%	70.60%	33.06%	16.05%	9.62%	6.35%	5.52%	0.00%
Alaska	100%	74.79%	25.21%	0.00%	1.93%	15.18%	1.84%	6.26%	0.00%
Arizona	100%	10.74%	87.09%	57.76%	18.43%	5.46%	4.00%	1.45%	2.17%
Arkansas	100%	0.00%	98.71%	29.35%	38.70%	16.35%	6.94%	7.37%	1.29%
California	100%	5.61%	58.90%	31.82%	15.05%	8.47%	1.76%	1.79%	35.49%
Colorado	100%	0.22%	67.37%	32.17%	18.43%	11.88%	3.00%	1.89%	32.41%
Connecticut	100%	0.00%	82.65%	20.89%	29.01%	21.87%	9.63%	1.14%	17.35%
Delaware	100%	75.14%	24.86%	0.00%	18.78%	3.42%	1.09%	1.57%	0.00%
District of Columbia	100%	0.00%	100.0%	100.00%	0.00%	0.00%	0.00%	0.00%	0.00%
Florida	100%	14.96%	84.14%	38.00%	23.87%	12.50%	4.50%	5.27%	0.90%
Georgia	100%	25.42%	74.58%	36.64%	10.15%	10.06%	3.04%	14.69%	0.00%
Hawaii	100%	11.80%	88.20%	88.20%	0.00%	0.00%	0.00%	0.00%	0.00%
Idaho	100%	1.69%	89.03%	0.00%	36.78%	22.16%	7.70%	22.35%	9.28%
Illinois	100%	0.67%	42.95%	13.59%	11.96%	9.12%	4.42%	3.86%	56.38%
Indiana	100%	0.11%	99.49%	27.01%	37.44%	21.14%	6.15%	7.76%	0.40%
Iowa	100%	0.00%	99.11%	19.08%	48.11%	10.14%	10.52%	11.26%	0.89%
Kansas	100%	0.09%	99.83%	57.84%	10.25%	17.42%	7.35%	6.97%	0.08%
Kentucky	100%	8.30%	84.83%	49.46%	4.63%	8.94%	14.76%	7.03%	6.86%
Louisiana	100%	9.31%	89.74%	57.41%	17.03%	5.73%	6.84%	2.72%	0.96%
Maine	100%	0.00%	80.70%	0.00%	10.65%	25.51%	33.34%	11.20%	'19.30%
Maryland	100%	26.93%	26.31%	19.36%	3.39%	0.96%	1.88%	0.73%	46.75%
Massachusetts	100%	0.00%	99.43%	14.93%	36.63%	36.06%	5.95%	5.86%	0.57%
Michigan	100%	47.00%	51.20%	29.15%	14.04%	2.44%	3.28%	2.29%	1.80%
Minnesota	100%	0.06%	77.82%	20.38%	24.44%	13.80%	7.58%	11.62%	22.12%
Mississippi	100%	0.30%	99.70%	5.02%	69.17%	11.33%	5.97%	8.21%	0.00%

Exhibit III-53 (Cont'd)

Total Sewerage Expenditures -- 1972 -- % Distribution by Governmental Unit and State

	Total	Counties	Municipalities (all sizes)	Municipality Size Group (000s)					Special District
				100+	25-100	10-25	5-10	<5	
Missouri	100%	0.05%	65.72%	32.18%	9.29%	10.61%	6.03%	7.62%	34.22%
Montana	100%	21.68%	77.29%	0.00%	45.69%	10.76%	3.84%	17.00%	1.03%
Nebraska	100%	0.00%	89.90%	56.13%	9.08%	14.18%	4.24%	6.27%	10.10%
Nevada	100%	0.69%	61.25%	13.38%	25.75%	18.40%	2.07%	1.66%	38.06%
New Hampshire	100%	0.10%	99.65%	-0.00%	49.59%	19.38%	20.71%	9.99%	0.25%
New Jersey	100%	0.94%	48.01%	6.88%	11.55%	10.47%	14.19%	4.91%	51.05%
New Mexico	100%	2.65%	97.35%	67.53%	13.42%	7.91%	4.49%	4.01%	0.00%
New York	100%	33.19%	66.81%	38.77%	12.28%	8.96%	4.63%	2.18%	0.00%
North Carolina	100%	12.26%	84.72%	32.31%	28.11%	12.43%	5.93%	5.94%	3.01%
North Dakota	100%	0.07%	99.93%	0.00%	89.39%	4.70%	3.77%	8.06%	0.00%
Ohio	100%	17.58%	81.89%	38.22%	22.25%	12.70%	5.33%	3.39%	0.53%
Oklahoma	100%	0.00%	98.06%	51.08%	15.93%	12.93%	6.49%	11.62%	1.94%
Oregon	100%	7.29%	83.22%	30.75%	16.86%	16.73%	10.92%	7.96%	9.49%
Pennsylvania	100%	0.84%	35.95%	13.38%	7.54%	6.13%	5.51%	3.39%	63.20%
Rhode Island	100%	0.00%	58.50%	10.15%	33.70%	13.88%	0.46%	0.31%	41.50%
South Carolina	100%	3.79%	63.38%	31.21%	11.28%	6.13%	11.51%	3.25%	32.84%
South Dakota	100%	0.00%	92.77%	0.00%	52.28%	14.04%	17.90%	8.55%	7.23%
Tennessee	100%	0.13%	97.67%	67.28%	4.21%	12.35%	6.36%	7.48%	2.20%
Texas	100%	0.00%	85.13%	49.96%	16.26%	10.06%	4.07%	4.79%	14.87%
Utah	100%	3.56%	51.27%	8.36%	14.04%	10.19%	11.49%	7.21%	45.16%
Vermont	100%	0.00%	100.0%	0.00%	15.28%	17.46%	45.88%	21.39%	0.00%
Virginia	100%	37.34%	50.60%	34.07%	5.98%	5.38%	3.17%	2.01%	12.06%
Washington	100%	0.30%	68.06%	46.46%	5.47%	10.47%	8.65%	3.02%	31.64%
West Virginia	100%	0.00%	84.78%	0.00%	41.31%	18.71%	8.67%	16.08%	15.22%
Wisconsin	100%	0.03%	65.79%	16.05%	19.18%	19.23%	6.18%	5.14%	34.18%
Wyoming	100%	0.00%	90.92%	0.00%	53.01%	13.29%	12.80%	11.81%	9.08%

129

Exhibit III-54

State Contributions Towards the Construction of Publicly-owned Treatment Facilities, In Percentage Terms

State	EPA ESTIMATES	State	EPA ESTIMATES
Alabama	0.0%	Montana	0.0%
Alaska	12.5%	Nebraska	12.5%
Arizona	5.0%	Nevada	0.0%
Arkansas	0.0%	New Hampshire	20.0%
California	12.5%	New Jersey	15.0%
Colorado	5.0%	New Mexico	12.5%
Connecticut	15.0%	New York	12.5%
Delaware	10.0%	North Carolina	12.5%
District of Columbia	0.00%	North Dakota	0.00%
Florida	0.00%	Ohio	0.00%
Georgia	0.00%	Oklahoma	0.00%
Hawaii	10.00%	Oregon	0.00%
Idaho	15.0%	Pennsylvania	0.00%
Illinois	0.00%	Rhode Island	15.0%
Indiana	10.0%	South Carolina	0.00
Iowa	0.00%	South Dakota	5.0%
Kansas	0.00%	Tennessee	0.00%
Kentucky	0.00%	Texas	0.00%
Louisiana	0.00%	Utah	0.00%
Maine	10.0%	Vermont	15.0%
Maryland	12.5%	Virginia	10.0%
Massachusetts	15.0%	Washington	15.0%
Michigan	5.0%	West Virginia	0.0%
Minnesota	15.0%	Wisconsin	5.0%
Mississippi	0.0%	Wyoming	0.0%
Missouri	15.0%		

SOURCE: Meta Systems, Inc., Evaluation of Methods for Financing Municipal Waste Treatment Works, EPA 60015-75-001, Washington, D.C., February 1975.

is dependent on their size; smaller communities typically have lower ratings and must bear higher interest rates.

Different interest rates are, therefore, assumed for the different types/sizes of government. We are further assuming that interest rates and presented rates do not vary between states.

Exhibit III-55	
Assumed Interest Cost on State	
and Local Government Debt	
Type of Government	Interest Rates
State Government	6.25%
County Government	6.55%
Municipalities:	
population over 100,000	6.55%
population 25-100,000	6.70%
population 10-25,000	6.85%
population 5-10,000	7.00%
population under 5,000	7.15%
Special Districts	6.75%

The proportion of annual expenditures financed from accumulated surplus may be treated in a manner analogous to the proportion financed from bond issues, assuming that the surplus will be replenished and that any existing surplus earns r% interest. In other words, AC_{ij} refers to both the amount financed from accumulated surplus and the amount financed from bond issues.

As expenditures are assumed to occur annually, for any given year, the amount of funds required for the sinking fund must be based on the previous expenditures. As expenditures are assumed to take place in equal annual increments, this simply involves multiplying AC_{ij} by the number of years elapsed since the beginning of the P.L. 92-500 construction program, or January 1, 1973.

Baseline Expenditures: Baseline expenditures have been defined as the expenditures which would have taken place in the absence of the Act.

One estimate of baseline expenditures used in this study was zero. This does not indicate that no expenditures would have been undertaken without the Act; rather the incidence calculations using zero baseline indicate the distribution of the total burden of water pollution abatement under P.L. 92-500.

The National Commission on Water Quality extrapolated past trends in water pollution abatement expenditures; their calculations yielded the high and low baselines presented in Exhibit III-3. The inclusion of baselines implies that differential (as opposed to total) incidence is being calculated. In our calcu-

131

lations, we have used both a zero baseline and the Commission's high baselines.

For the high baseline, the federal share was assumed to be 15.9%, the state shares were assumed to correspond to the current shares presented in Exhibit III-54. The method of financing adopted by the various levels and types of governments was assumed to be the same for both baseline costs and costs incurred under the Act.

Operations and Maintenance Costs: O&M costs have been estimated for the different categories of "needs" by Metcalf & Eddy. For categories I, II, and IVB, this was 2.43% of the total capital cost, and for categories I-V, it was 1.48%. The annual O&M costs for any given year depend on the amount of capital in place. According to our assumptions, this may be expressed as

$$OM_{ijt} = \frac{.0243 \times TC_{ij}}{10} t$$

for categories I, II, and IVB, and

$$OM_{ijt} = \frac{.0148 \times TC_{ij}}{10} t$$

for categories I-V, and

$$OM_{ijt} = \frac{.243 \times TBC_{ij}}{10} t$$

for the high baseline costs, where

t is the number of years elapsed since 1973;
TC is the total cost of abatement; and
TBC is the local baseline cost.

For both baseline expenditures and costs incurred under P.L. 92-500, we assume that the federal government does not contribute toward O&M costs at all. As this has been the case in the past and is at the present time, it appears a plausible assumption.

With respect to state contributions, only New York State makes grants towards operations and maintenance expenditures, in the amount of 33-1/3%. In other words, all O&M costs are assumed to be paid entirely out of local sources, with the exception of New York.

The Method of Financing for Capital Costs (Matrix V)

We are assuming that the method of financing for both base-line expenditures and costs incurred under the current legislation

132

is the same. Financing methods will be selected at the federal, state and local levels.

At the federal and state level, it was assumed that pollution control expenditures are financed out of tax increases exclusively. At the local level, expenditure reductions are considered as well. The calculations described below attempted to establish the relative importance of each revenue source in financing water pollution abatement expenditure.

Financing the Federal Share: Federal expenditures on pollution control are assumed to be financed out of general revenues, with the exception of social security taxes, which are assumed earmarked.

The percentage distribution of federal revenues by source is depicted in Exhibit III-56 for 1975. We are assuming that the distribution excluding social security taxes is relevant for financing the federal share of abatement expenditures.

For 1977, 1980, and 1985, shifts in the composition of the budget receipts have been estimated on the basis of income elasticities of the various taxes amd GNP forecasts for those years.

GNP forecasts were generated by CONSAD. These were $1,155.2 billion for 1972, $1,825.0 billion for 1977, and $2,460.0 billion for 1980. The income elasticities of the various taxes have been calculated on the basis of the observed relationship between tax shares and the GNP growth rate. The relevant calculations are described below.

Exhibit III-56

Federal Budget Receipts by Source, 1975*

SOURCE	$Millions	% Distribution	% Distribution Excluding Social Security Taxes
Individual Income Tax	$117,700	42.2%	61.1%
Corporation Income Tax	38,500	13.8%	20.0%
Social Security Tax	86,225	30.9%	--
Excise Tax	19,947	7.2%	10.4%
Estate and Gift Tax	4,800	1.7%	2.5%
Customs Duties	3,910	1.4%	2.0%
Miscellaneous	7,668	2.8%	4.0%
TOTAL	$278,750	100.0%	100.0%

*1975 Estimates

SOURCE: The Budget of the United States Government, Fiscal Year 1976, Table #11, p. 332.

133

$$Q_{it} = \frac{T_{it}}{\sum_1 T_{it}}$$

where

Q_{it} = share of tax i in (relevant) total tax revenues in year t.

T_{it} = tax revenues of type i in year t.

and

Y_t = GNP in year t

r = annual GNP growth rate

α_i = GNP elasticity of tax revenue i.

$$\Delta Y_t = [(1+r)^t - 1]$$

Let $h_i = \Delta T_{it} = \alpha_i \Delta Y_t$

$$= \alpha_i [(1+r)^t - 1]$$

Then $T_{it} = (1+h_i) T_{io}$

$$= [1 - \alpha_i + \alpha_i (1+r)^t] T_{io}$$

and $Q_{it} = \dfrac{[1 - \alpha_i + \alpha_i (1+r)^t] T_{io}}{\sum_j [1 - \alpha_i + \alpha_i (1+r)^t] T_{jo}}$

According to this methodology, empirical tax elasticities have been calculated for the personal income tax, the corporation net income tax, the excise tax, the estate and gift tax, and customs and duties. Two alternative sets of elasticities were calculated, one on the basis of budgetary trends between 1965 and 1972, and one on the basis of trends between 1965 and 1975.

The change in GNP was 68.67% between 1965 and 1972, and 107% between 1965 and 1975. As the observed income elasticities based on the two periods varied substantially, comparisons were made with other public finance studies such as the Tax Foundation, and Dew and Friedlander. The elasticity estimates selected for this study were based, in part, on these comparisons. These estimates, as well as the elasticities selected for the purposes of this study, are presented in Exhibit III-57.

Using the observed income elasticities, α_i, and the projected GNP growth, increases for each of the taxes were estimated, and their respective budget shares were calculated for 1977, 1980, and 1985. These are presented in Exhibit III-58.

Exhibit III-57

Empirical GNP Elasticities of Selected Taxes for the

Federal Government

Tax	Time Period 1965-72	1965-75	Elasticity Used in Study
Personal Income Tax	1.37%	1.32%	1.35%
Corporation Net Income Tax	0.38%	0.48%	0.50%
Excise Tax	0.09%	0.34%	0.35%
Estate & Gift Tax	1.46%	0.72%	1.00%
Customs Duties	1.98%	1.68%	1.70%

Exhibit III-58

The Projected Composition of Federal Revenues

	1971	1980	1985
Personal Income Tax	69.1%	71.9%	74.5%
Corporation Income Tax	17.1%	15.2%	13.5%
Excise Tax	7.6%	6.5%	5.5%
Estate & Gift Tax	3.5%	3.5%	3.4%
Customs and Duties	2.7 %	2.9%	3.1%

Financing the State Shares: We are assuming that states finance
their share of pollution control expenditures out of general
revenues, subject to some restrictions. The composition of general
revenues or tax shares were projected on the basis of observed
tax GNP elasticities. The GNP elasticities of state taxes were
calculated according to the methodology used for federal tax
elasticities.

Not all tax revenues are available for financing pollution
control expenditures because of earmarking; a certain proportion
of several taxes is earmarked for special purposes, and, therefore,
is not available for financing the state share of pollution control
expenditures. The earmarked proportion of the various taxes have
been studied by the Tax Foundation in 1965, for each state; their
findings are presented in Exhibit III-59.

The projected tax revenue structures were then adjusted to
allow for earmarking, and the resulting proportions were normal-
ized. The projected distribution of revenues are presented in
Exhibits III-60 through III-62 for 1977, 1980, and 1985.

Financing at the Local Level: Local governments can finance their
expenditures on pollution abatement from increased tax/charge revenues or from expenditure reductions. In this study, we have

135

Exhibit III-59

Assumed Percentages of State Government Taxes Which Are Available to

Finance Intergovernmental Transfers for Sewerage, by States

STATE	General Sales and Gross Receipts	Selective Sales	Individual Income	Corporation Net Income	Property	Severance	Death and Gift
Alabama	8 %	18 %	0 %	0 %	38 %	64 %	100 %
Alaska	-	57 %	100 %	100 %	--	98 %	100 %
Arizona	27 %	32 %	100 %	100 %	100 %	--	100 %
Arkansas	100 %	44 %	100 %	100 %	100 %	63 %	100 %
California	100 %	46 %	100 %	100 %	2 %	0 %	100 %
Colorado	40 %	24 %	100 %	100 %	0 %	0 %	100 %
Connecticut	100 %	61 %	100 %	100 %	100 %	--	100 %
Delaware	100 %	100 %	100 %	100 %	100 %	--	100 %
Florida	87 %	53 %	--	100 %	48 %	100 %	100 %
Georgia	100 %	44 %	100 %	100 %	100 %	--	100 %
Hawaii	100 %	68 %	100 %	100 %	--	--	100 %
Idaho	84 %	29 %	91 %	100 %	100 %	100 %	100 %
Illinois	71 %	60 %	100 %	100 %	100 %	100 %	100 %
Indiana	100 %	31 %	100 %	100 %	10 %	100 %	100 %
Iowa	84 %	39 %	100 %	100 %	0 %	--	100 %
Kansas	100 %	30 %	100 %	100 %	0 %	0 %	95 %
Kentucky	100 %	49 %	100 %	100 %	100 %	100 %	100 %
Louisiana	0 %	29 %	0 %	0 %	52 %	5 %	100 %
Maine	100 %	52 %	100 %	100 %	65 %	--	100 %
Maryland	100 %	52 %	75 %	100 %	100 %	--	100 %
Massachusetts	100 %	64 %	51 %	73 %	100 %	--	100 %
Michigan	43 %	41 %	100 %	100 %	100 %	100 %	100 %
Minnesota	75 %	48 %	7 %	7 %	50 %	48 %	82 %
Mississippi	82 %	37 %	100 %	100 %	100 %	70 %	100 %
Missouri	100 %	21 %	100 %	100 %	56 %	--	100 %
Montana	--	34 %	75 %	75 %	22 %	97 %	100 %
Nebraska	100 %	35 %	100 %	100 %	94 %	0 %	100 %

Exhibit III-59 (Cont'd)

Assumed Percentages of State Government Taxes Which Are Available To
Finance Intergovernmental Transfers for Sewerage, by States

STATE	General Sales and Gross Receipts	Selective Sales	Individual Income	Corporation Net Income	Property	Severance	Death and Gift
Nevada	100 %	69 %	--	--	100%	100%	--
New Hampshire	100 %	61 %	0 %	0 %	100%	0%	100%
New Jersey	100 %	100 %	0 %	100 %	100%	--	100%
New Mexico	100 %	42 %	100 %	100 %	90%	97%	100%
New York	80 %	81 %	89 %	100 %	100%	--	100%
North Carolina	100 %	40 %	100 %	100 %	35%	34%	100%
North Dakota	100 %	38 %	100 %	100 %	10%	100%	35%
Ohio	100 %	47 %	100 %	100 %	35%	63%	100%
Oklahoma	5 %	52 %	100 %	100 %	--	18%	100%
Oregon	100 %	34 %	100 %	100 %	100%	--	100%
Pennsylvania	0 %	56 %	100 %	100 %	100%	--	100%
Rhode Island	100 %	92 %	100 %	100 %	--	--	100%
South Carolina	0 %	38 %	100 %	97 %	100%	100%	100%
South Dakota	100 %	47 %	--	3 %	--	--	100%
Tennessee	2 %	21 %	73 %	100 %	--	33%	100%
Texas	48 %	53 %	--	--	0%	100%	100%
Utah	87 %	22 %	0 %	0 %	0%	100%	100%
Vermont	100 %	65 %	100 %	100 %	90%	--	100%
Virginia	--	49 %	100 %	100 %	99%	0%	100%
Washington	100 %	47 %	--	--	35%	--	100%
West Virginia	97 %	42 %	100 %	100 %	0%	--	100%
Wisconsin	1 %	40 %	54 %	51 %	25%	20%	100%
Wyoming	100 %	11 %	--	--	14%	100%	100%

SOURCE: Based on data in Tax Foundation, Inc., Earmarked State Taxes, June 1965.

Exhibit III-60

Percentage Distribution of Certain Projected State Government Tax Revenues in 1977

State	General Sales Tax	Selective Sales Tax	Individual Income Tax	Corporation Income Tax	Property Tax	Death & Gift Tax	Severance Tax	Total
Alabama	33.33%	36.67%	21.42%	5.17%	2.37%	0.69%	0.35%	100%
Alaska	0.00%	20.78%	47.10%	8.46%	0.00%	0.05%	23.60%	100%
Arizona	37.21%	23.01%	22.14%	6.58%	9.74%	1.32%	0.00%	100%
Arkansas	31.76%	33.18%	25.54%	8.37%	0.26%	0.24%	0.64%	100%
California	28.54%	16.70%	39.99%	8.37%	2.99%	3.38%	0.02%	100%
Colorado	38.46%	21.13%	32.81%	4.66%	0.21%	2.68%	0.05%	100%
Connecticut	44.19%	30.45%	7.71%	13.55%	0.00%	4.11%	0.00%	100%
Delaware	0.00%	36.69%	52.16%	9.44%	0.14%	1.56%	0.00%	100%
District of Columbia	0.00%	0.00%	0.00%	0.00%	0.00%	0.00%	0.00%	100%
Flordia	62.00%	29.52%	0.00%	1.42%	4.69%	2.22%	0.16%	100%
Georgia	33.45%	28.30%	28.44%	8.71%	0.33%	0.77%	0.00%	100%
Hawaii	45.70%	12.34%	38.50%	2.43%	0.00%	1.03%	0.00%	100%
Idaho	34.73%	25.39%	27.37%	7.42%	0.26%	4.76%	0.08%	100%
Illinois	31.40%	26.46%	34.56%	5.91%	0.08%	1.60%	0.00%	100%
Indiana	33.88%	27.09%	34.15%	1.05%	2.34%	1.48%	0.01%	100%
Iowa	28.58%	20.57%	40.81%	7.49%	0.01%	2.54%	0.00%	100%
Kansas	34.37%	27.48%	25.75%	9.33%	1.44%	1.53%	0.08%	100%
Kentucky	41.98%	22.04%	25.38%	5.88%	2.25%	1.45%	1.02%	100%
Louisiana	31.26%	22.64%	16.15%	9.70%	2.21%	0.81%	17.23%	100%
Maine	46.34%	32.15%	13.62%	3.44%	2.26%	2.19%	0.00%	100%
Maryland	23.27%	18.54%	48.04%	7.24%	2.34%	0.57%	0.00%	100%
Massachusetts	10.23%	21.15%	48.86%	16.76%	0.01%	3.00%	0.00%	100%
Michigan	30.67%	19.31%	35.09%	11.31%	2.39%	1.20%	0.02%	100%
Minnesota	21.41%	19.56%	46.90%	9.26%	0.11%	1.94%	0.82%	100%
Mississippi	58.06%	23.80%	13.19%	2.69%	0.37%	0.55%	1.35%	100%

138

Exhibit III-60 (Cont'd)

Percentage Distribution of Certain Projected State Government Tax Revenues in 1977

State	General Sales Tax	Selective Sales Tax	Individual Income Tax	Corporation Income Tax	Property Tax	Death & Gift Tax	Severance Tax	Total
Missouri	36.43 %	20.22 %	32.06 %	7.08 %	0.20 %	2.01 %	0.00 %	100 %
Montana	0.00 %	28.13 %	59.27 %	6.35 %	2.35 %	2.20 %	1.70 %	100 %
Nebraska	38.93 %	33.72 %	25.00 %	1.85 %	0.09 %	0.26 %	0.12 %	100 %
Nevada	39.12 %	51.62 %	0.00 %	0.00 %	9.19 %	0.00 %	0.06 %	100 %
New Hampshire	0.00 %	71.84 %	7.31 %	14.52 %	3.34 %	2.96 %	0.03 %	100 %
New Jersey	44.11 %	29.20 %	2.01 %	12.46 %	6.47 %	5.74 %	0.00 %	100 %
New Mexico	50.72 %	20.69 %	15.64 %	2.01 %	2.97 %	0.61 %	7.37 %	100 %
New York	27.72 %	17.60 %	39.41 %	12.51 %	0.29 %	2.47 %	0.00 %	100 %
North Carolina	21.82 %	33.33 %	33.87 %	7.52 %	1.71 %	1.78 %	0.00 %	100 %
North Dakota	54.86 %	20.65 %	13.62 %	8.59 %	0.49 %	0.66 %	1.13 %	100 %
Ohio	44.63 %	34.54 %	8.42 %	8.44 %	2.66 %	1.28 %	0.03 %	100 %
Oklahoma	17.80 %	37.61 %	23.79 %	4.48 %	0.00 %	3.70 %	12.61 %	100 %
Oregon	0.00 %	27.94 %	61.64 %	7.52 %	0.01 %	2.13 %	0.76 %	100 %
Pennsylvania	25.97 %	25.56 %	27.73 %	16.11 %	1.56 %	3.07 %	0.00 %	100 %
Rhode Island	36.17 %	21.62 %	28.67 %	10.31 %	0.00 %	3.24 %	0.00 %	100 %
South Carolina	40.95 %	23.49 %	26.58 %	7.83 %	0.21 %	0.93 %	0.00 %	100 %
South Dakota	63.62 %	33.37 %	0.00 %	0.54 %	0.00 %	2.47 %	0.00 %	100 %
Tennessee	51.52 %	39.59 %	1.46 %	12.64 %	0.00 %	4.80 %	0.00 %	100 %
Texas	44.02 %	41.71 %	0.00 %	0.00 %	1.98 %	1.64 %	10.65 %	100 %
Utah	42.15 %	16.32 %	34.26 %	2.92 %	2.95 %	0.72 %	0.68 %	100 %
Vermont	15.58 %	39.57 %	38.06 %	4.24 %	0.11 %	2.45 %	0.00 %	100 %
Virginia	24.39 %	24.48 %	42.56 %	6.37 %	0.64 %	1.53 %	0.04 %	100 %
Washington	51.74 %	27.85 %	0.00 %	0.00 %	17.10 %	3.31 %	0.00 %	100 %
West Virginia	41.77 %	30.28 %	24.48 %	2.41 %	0.08 %	0.97 %	0.00 %	100 %
Wisconsin	24.43 %	15.41 %	48.20 %	5.74 %	4.55 %	1.67 %	0.01 %	100 %
Wyoming	51.23 %	31.05 %	0.00 %	0.00 %	6.91 %	1.24 %	9.57 %	100 %

139

Exhibit III-61

Percentage Distribution of Certain Projected State Government Tax Revenues In 1980

State	General Sales Tax	Selective Sales Tax	Individual Income Tax	Corporation Income Tax	Property Tax	Death & Gift Tax	Severance	Total
Alabama	32.95%	35.28%	23.22%	5.42%	2.04%	0.77%	0.33%	100%
Alaska	0.00%	18.74%	47.25%	8.68%	0.00%	0.04%	25.29%	100%
Arizona	36.40%	22.54%	23.60%	7.02%	9.03%	1.41%	0.00%	100%
Arkansas	30.50%	31.74%	28.26%	8.57%	0.26%	0.22%	0.44%	100%
California	27.63%	15.30%	43.30%	7.76%	2.64%	3.34%	0.02%	100%
Colorado	39.86%	20.08%	33.21%	4.05%	0.14%	2.62%	0.03%	100%
Connecticut	45.73%	28.90%	9.03%	13.61%	0.00%	3.73%	0.00%	100%
Delaware	0.00%	37.32%	52.12%	9.21%	0.12%	1.23%	0.00%	100%
District of Columbia	0.00%	0.00%	0.00%	0.00%	0.00%	0.00%	0.00%	100%
Florida	64.22%	27.29%	0.00%	1.35%	4.67%	2.30%	0.17%	100%
Georgia	32.32%	26.96%	30.60%	8.98%	0.32%	0.82%	0.00%	100%
Hawaii	44.80%	11.43%	40.61%	2.11%	0.00%	1.05%	0.00%	100%
Idaho	36.35%	23.97%	26.67%	7.37%	0.18%	5.38%	0.08%	100%
Illinois	29.94%	25.70%	36.85%	5.99%	0.07%	1.45%	0.00%	100%
Indiana	31.91%	26.00%	37.09%	1.08%	2.45%	1.45%	0.01%	100%
Iowa	27.26%	18.61%	43.71%	8.02%	0.01%	2.39%	0.00%	100%
Kansas	33.14%	26.38%	27.77%	10.17%	1.11%	1.37%	0.06%	100%
Kentucky	42.77%	20.06%	27.20%	5.64%	1.86%	1.36%	1.10%	100%
Louisiana	32.76%	20.99%	18.24%	10.39%	1.99%	0.75%	14.88%	100%
Maine	47.98%	30.04%	14.34%	3.44%	2.24%	1.96%	0.00%	100%
Maryland	22.88%	16.63%	50.46%	7.41%	2.15%	0.47%	0.00%	100%
Massachusetts	9.88%	19.47%	50.31%	17.49%	0.00%	2.86%	0.00%	100%
Michigan	28.74%	18.58%	37.86%	11.60%	2.04%	1.17%	0.01%	100%
Minnesota	21.23%	17.91%	49.03%	9.28%	0.07%	1.89%	0.59%	100%
Mississippi	60.00%	21.83%	14.12%	2.26%	0.25%	0.54%	1.00%	100%

140

Exhibit III-61 (Cont'd)

Percentage Distribution of Certain Projected State Government Tax Revenues In 1980

State	General Sales Tax	Selective State Tax	Individual Income Tax	Corporation Income Tax	Property Tax	Death & Gift Tax	Severance Tax	Total
Missouri	38.14%	18.66%	33.46%	7.61%	0.14%	2.00%	0.00%	100%
Montana	0.00%	24.81%	64.13%	6.06%	1.64%	1.99%	1.37%	100%
Nebraska	39.94%	31.56%	26.85%	1.26%	0.06%	0.25%	0.08%	100%
Nevada	39.79%	50.58%	0.00%	0.00%	9.57%	0.00%	0.06%	100%
New Hampshire	0.00%	71.83%	7.79%	14.61%	3.19%	2.56%	0.02%	100%
New Jersey	44.73%	26.57%	2.11%	13.65%	7.09%	5.84%	0.00%	100%
New Mexico	53.73%	19.20%	16.37%	1.36%	2.45%	0.64%	6.24%	100%
New York	29.00%	16.40%	39.35%	12.58%	0.31%	2.36%	0.00%	100%
North Carolina	20.97%	32.77%	35.95%	6.95%	1.56%	1.79%	0.00%	100%
North Dakota	57.69%	17.99%	13.42%	9.17%	0.32%	0.64%	0.76%	100%
Ohio	46.30%	31.91%	9.31%	8.88%	2.33%	1.24%	0.02%	100%
Oklahoma	16.78%	36.40%	26.33%	4.26%	0.00%	3.84%	12.39%	100%
Oregon	0.00%	27.77%	62.73%	6.84%	0.01%	1.82%	0.83%	100%
Pennsylvania	24.89%	24.24%	29.63%	16.66%	1.73%	2.86%	0.00%	100%
Rhode Island	37.29%	18.96%	30.17%	10.33%	0.00%	3.25%	0.00%	100%
South Carolina	41.71%	20.94%	28.54%	7.73%	0.18%	0.90%	0.00%	100%
South Dakota	67.36%	29.66%	0.00%	0.48%	0.00%	2.51%	0.00%	100%
Tennessee	52.69%	27.48%	1.34%	13.32%	0.00%	5.17%	0.00%	100%
Texas	46.49%	40.87%	0.00%	0.00%	1.68%	1.59%	9.38%	100%
Utah	42.51%	14.79%	36.78%	2.48%	2.32%	0.65%	0.48%	100%
Vermont	15.58%	38.40%	39.48%	4.09%	0.07%	2.38%	0.00%	100%
Virginia	24.49%	22.20%	45.19%	6.12%	0.43%	1.54%	0.04%	100%
Washington	50.16%	27.29%	0.00%	0.00%	19.04%	3.51%	0.00%	100%
West Virginia	40.42%	29.35%	36.73%	2.51%	0.07%	0.92%	0.00%	100%
Wisconsin	24.22%	14.01%	50.82%	5.19%	4.23%	1.52%	0.01%	100%
Wyoming	52.90%	29.98%	0.00%	0.00%	5.20%	1.22%	10.70%	100%

Exhibit III-62

Percentage Distribution of Certain Projected State Government Tax Revenues in 1985

State	General Sales Tax	Selective Sales Tax	Individual Income Tax	Corporation Income Tax	Property Tax	Death & Gift Tax	Severance Tax	Total
Alabama	32.65%	34.14%	24.69%	5.62%	1.76%	0.82%	0.32%	100%
Alaska	0.00%	17.16%	47.37%	8.85%	0.00%	0.02%	26.59%	100%
Arizona	35.78%	22.18%	24.72%	7.35%	8.49%	1.47%	0.00%	100%
Arkansas	29.47%	30.56%	30.49%	8.74%	0.26%	0.19%	0.29%	100%
California	26.91%	14.20%	45.91%	7.28%	2.36%	3.32%	0.02%	100%
Colorado	40.99%	19.24%	33.52%	3.56%	0.09%	2.58%	0.02%	100%
Connecticut	46.93%	27.70%	8.27%	13.66%	0.00%	3.43%	0.00%	100%
Delaware	0.00%	37.84%	52.09%	9.02%	0.10%	0.94%	0.00%	100%
District of Columbia	0.00%	0.00%	0.00%	0.00%	0.00%	0.00%	0.00%	100%
Florida	65.85%	25.67%	0.00%	1.30%	4.65%	2.36%	0.17%	100%
Georgia	31.44%	25.92%	32.29%	9.19%	0.31%	0.85%	0.00%	100%
Hawaii	44.10%	10.72%	42.26%	1.85%	0.00%	1.07%	0.00%	100%
Idaho	37.73%	22.76%	26.07%	7.34%	0.11%	5.91%	0.07%	100%
Illinois	28.76%	25.08%	38.71%	6.05%	0.07%	1.33%	0.00%	100%
Indiana	30.27%	25.09%	39.56%	1.11%	2.55%	1.42%	0.00%	100%
Iowa	26.24%	17.10%	45.95%	8.43%	0.00%	2.27%	0.00%	100%
Kansas	32.08%	25.45%	29.48%	10.88%	0.82%	1.24%	0.04%	100%
Kentucky	43.39%	18.50%	28.63%	5.46%	1.56%	1.30%	1.17%	100%
Louisiana	34.03%	19.59%	20.02%	10.98%	1.80%	0.70%	12.88%	100%
Maine	49.27%	28.37%	14.92%	3.45%	2.22%	1.77%	0.00%	100%
Maryland	22.59%	15.20%	52.27%	7.54%	2.00%	0.40%	0.00%	100%
Massachusetts	9.62%	18.23%	51.38%	18.02%	0.00%	2.75%	0.00%	100%
Michigan	27.13%	17.98%	40.16%	11.84%	1.75%	1.14%	0.01%	100%
Minnesota	21.08%	16.63%	50.68%	9.30%	0.04%	1.85%	0.41%	100%
Mississippi	61.51%	20.31%	14.84%	1.94%	0.16%	0.52%	0.73%	100%

Exhibit III-62 (Cont'd)

Percentage Distribution of Certain Projected State Government Tax Revenues in 1985

State	General Sales Tax	Selective Sales Tax	Individual Income Tax	Corporation Income Tax	Property Tax	Death & Gift Tax	Severance Tax	Total
Missouri	37.91 %	17.42 %	34.58 %	8.03 %	0.08 %	1.99 %	0.00 %	100 %
Montana	0.00 %	22.19 %	67.95 %	5.84 %	1.08 %	1.83 %	1.10 %	100 %
Nebraska	40.77 %	29.76 %	28.36 %	0.78 %	0.04 %	0.23 %	0.05 %	100 %
Nevada	40.28 %	49.82 %	0.00 %	0.00 %	9.85 %	0.00 %	0.06 %	100 %
New Hampshire	0.00 %	71.82 %	8.17 %	14.68 %	3.08 %	2.24 %	0.01 %	100 %
New Jersey	45.24 %	24.44 %	2.20 %	14.61 %	7.59 %	5.92 %	0.00 %	100 %
New Mexico	56.19 %	18.00 %	16.97 %	0.84 %	2.04 %	0.66 %	5.31 %	100 %
New York	30.09 %	15.38 %	39.29 %	12.64 %	0.33 %	2.26 %	0.00 %	100 %
North Carolina	20.29 %	32.34 %	37.62 %	6.50 %	1.45 %	1.80 %	0.00 %	100 %
North Dakota	59.87 %	15.94 %	13.27 %	9.62 %	0.19 %	0.63 %	0.48 %	100 %
Ohio	47.75 %	29.63 %	10.09 %	9.26 %	2.05 %	1.21 %	0.01 %	100 %
Oklahoma	15.68 %	35.30 %	28.63 %	4.06 %	0.00 %	3.96 %	12.19 %	100 %
Oregon	0.00 %	37.63 %	63.65 %	6.26 %	0.01 %	1.56 %	0.89 %	100 %
Pennsylvania	24.01 %	23.16 %	31.18 %	17.10 %	1.86 %	2.69 %	0.00 %	100 %
Rhode Island	38.17 %	16.85 %	31.37 %	10.35 %	0.00 %	3.26 %	0.00 %	100 %
South Carolina	42.31 %	18.94 %	30.07 %	7.65 %	0.15 %	0.89 %	0.00 %	100 %
South Dakota	70.43 %	26.60 %	0.00 %	0.43 %	0.00 %	2.54 %	0.00 %	100 %
Tennessee	53.64 %	25.79 %	1.24 %	13.86 %	0.00 %	5.47 %	0.00 %	100 %
Texas	48.62 %	40.13 %	0.00 %	0.00 %	1.42 %	1.54 %	8.28 %	100 %
Utah	42.30 %	13.59 %	38.74 %	2.14 %	1.82 %	0.59 %	0.32 %	100 %
Vermont	15.58 %	37.48 %	40.60 %	3.97 %	0.04 %	2.32 %	0.00 %	100 %
Virginia	24.58 %	20.40 %	47.27 %	5.92 %	0.26 %	1.55 %	0.03 %	100 %
Washington	48.75 %	26.79 %	0.00 %	0.00 %	20.77 %	3.69 %	0.00 %	100 %
West Virginia	39.28 %	28.56 %	28.64 %	2.58 %	0.06 %	0.87 %	0.00 %	100 %
Wisconsin	24.06 %	12.93 %	52.85 %	4.77 %	3.98 %	1.40 %	0.01 %	100 %
Wyoming	54.30 %	29.09 %	0.00 %	0.00 %	3.76 %	1.80 %	11.65 %	100 %

assumed that 90% of the expenditures will be financed from increased revenues, the remaining 10% from expenditure cuts. Expenditure reductions for all local governments and the reasons for
this assumption are discussed separately.

Pollution control expenditures at the local level may be undertaken by counties, special districts and municipalities. Local
governments typically finance their pollution control expenditures
from different revenue sources than state governments. Further,
the importance of the various revenue sources differs for the
three types of local governments. In the State of Illinois, for
example, we have assumed that the state government finances 34.6%
of its (capital) costs from the individual income tax, 31.4% from
the general sales tax, 26.5% from the selective sales tax, 5.9%
from the corporation income tax, and 1.6% from the death and gift
taxes. In contrast, counties in Illinois finance their costs
almost exclusively (82.4%) from the property tax, 7.6% from sales
taxes, and 10% from expenditure reductions. Special districts
raise 51.7% of their costs from the property tax, 38.3% from intergovernmental revenues from other local governments, and 10% from
expenditure reductions. Municipal governments in this state finance 50.4% of their pollution control costs from the property tax,
21.9% from the general sales tax, 8.5% from taxes on public utilities, 4.4% from motor vehicle licenses, 4.8% from other tax
revenues, and 10% from expenditure reductions. The importance of
these revenue sources for the four types of government are depicted in Exhibit III-63. Financing methods and assumptions will be
discussed for each type of local government: counties, special
districts, and municipalities.

For counties, the property tax is clearly the most important
source of revenue, as indicated by Exhibit III-64. In fact, in
approximately one-half of the states, over 90% of the county
governments' revenue comes from the property tax. The share of
the property tax in county government receipts has changed
appreciably between 1962 and 1972. However, forecasting on the
basis of these trends is somewhat difficult.

In the states where county governments have already reduced
the share of the property tax substantially, such reductions are
not likely to continue. On the other hand, county governments are
much more likely to broaden their tax base in states where this
has not yet occured. Therefore, in the absence of better information, we are assuming that the 1972 share of the property tax
will remain constant in 1983.

The remainder in each case will be divided between the general
sales tax--60%, the selective sales tax--14%, and the individual
income tax--17%. Because of the extreme variations in the
shares of the various revenue sources for the states, we are
assuming the uniform proportions described above.

144

Exhibit III-63

The Relative Importance of Revenue Sources for Different Types
of Government in Illinois

STATE   MUNICIPALITY

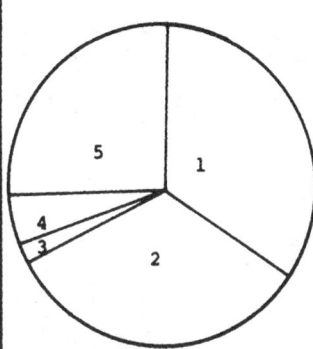 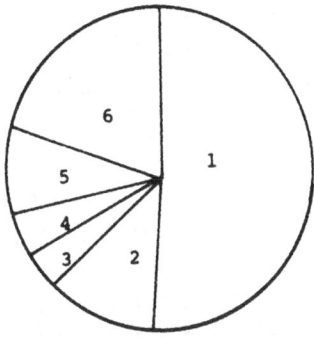

1. 34.6% individual income tax   1. 50.4% property tax
2. 31.4% general sales tax   2. 10.0% expenditure reduction
3. 1.6% death and gift taxes   3. 4.8% other tax revenues
4. 5.9% corporation income tax   4. 4.4% motor vehicle licenses
5. 26.5% selective sales tax   5. 8.5% taxes on public utilities
                                    6. 21.9% general sales tax

SPECIAL DISTRICT   COUNTY

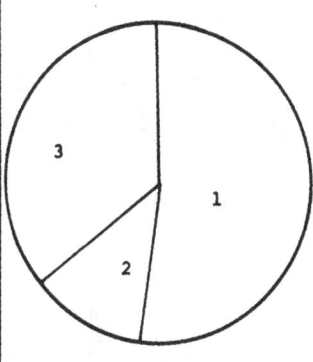 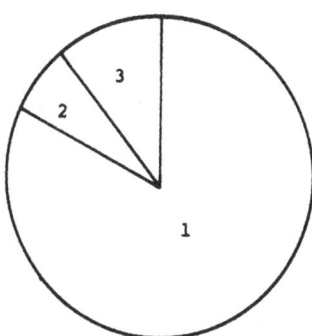

1. 51.7% property tax   1. 82.4% property tax
2. 10.0% expenditure reduction   2. 7.6% sales tax
3. 38.3% intergovernmental revenues 3. 10.0% expenditure reduction

145

Exhibit III-64

Property Tax as a Percent of All Tax Revenues for Counties And Special Districts (1972)

State	Percent Property Tax for Counties	for Special Districts	State	Percent Property Tax for Counties	for Special Districts
Alabama	49.22 %	27.30 %	Missouri	89.76 %	38.22 %
Alaska	84.15 %	27.30 %	Montana	94.41 %	20.29 %
Arizona	96.01 %	86.76 %	Nebraska	95.90 %	17.06 %
Arkansas	98.39 %	6.75 %	Nevada	53.19 %	29.39 %
California	94.65 %	47.70 %	New Hampshire	100.00 %	27.30 %
Colorado	95.97 %	11.90 %	New Jersey	99.68 %	0.00 %
Connecticut	0.00 %	2.38 %	New Mexico	93.19 %	27.30 %
Delaware	96.12 %	27.30 %	New York	64.12 %	27.30 %
District of Columbia	0.00 %	27.30 %	North Carolina	87.49 %	27.30 %
Florida	87.76 %	27.30 %	North Dakota	96.67 %	27.30 %
Georgia	92.68 %	27.30 %	Ohio	79.08 %	27.30 %
Hawaii	74.83 %	27.30 %	Oklahoma	99.58 %	27.30 %
Idaho	96.82 %	24.88 %	Oregon	95.69 %	23.53 %
Illinois	91.61 %	57.54 %	Pennsylvania	97.55 %	27.30 %
Indiana	89.81 %	74.07 %	Rhode Island	0.00 %	56.36 %
Iowa	97.73 %	48.23 %	South Carolina	97.14 %	65.57 %
Kansas	97.58 %	26.32 %	South Dakota	84.20 %	14.72 %
Kentucky	73.13 %	2.14 %	Tennessee	70.35 %	27.30 %
Louisiana	69.50 %	16.48 %	Texas	87.10 %	27.30 %
Maine	96.71 %	1.27 %	Utah	91.86 %	36.93 %
Maryland	66.86 %	27.30 %	Vermont	100.00 %	27.30 %
Massachusetts	98.37 %	27.30 %	Virginia	76.84 %	0.12 %
Michigan	96.63 %	27.30 %	Washington	72.30 %	0.36 %
Minnesota	99.26 %	27.30 %	West Virginia	98.85 %	27.30 %
Mississippi	83.07 %	27.30 %	Wisconsin	98.96 %	29.63 %
			Wyoming	99.11 %	3.77 %

In addition to the property tax, the general and selective
sales tax, and the individual income tax, counties may use accumulated surpluses for financing capital costs. These surpluses,
however, have been accumulated from general revenues--i.e., the
sources described above. Therefore, they do not need to be considered separately.

Special sewer districts have three main sources of revenue:
charges, property taxes, and intergovernmental revenue from other
local governments. Capital expenditures may also be financed from
accumulated surplus; however, assuming that the surplus has been
accumulated from general revenue, this does not need to be considered separately.

The Census of Governments gives the share of the property tax
in the budget of the special district, for some selected districts
in some states. Based on this information we have calculated the
property tax share for the given states. This is described in
Exhibit III-63. As the special districts described in the Census
account for most (approximately 80%) of all expenditures by sewer
districts, a national average has been calculated (at 27.3%) for
the property tax share, and applied to the 23 states for which no
information exists.

We have assumed that charges are used to finance the O&M
portion of the costs. Therefore, intergovernmental revenues,
receipts from other local governments, make up the balance. The
composition of receipts from other local governments was assumed
to be the same as the composition of municipal revenues in the
state in question.

Finally, we assume that the proportions remain constant over
time, as we feel that the data is inadequate for forecasting, for
the reasons described for county governments.

Municipal governments, according to the Census of Governments,
have the following sources of tax revenues for municipalities:
property tax, general sales tax, the selective sales tax, utility
taxes, motor vehicle licenses, income taxes, and others. Information is available on the magnitude and distribution of these
revenue sources for 1962 and 1972 from the Census of Governments.

For municipalities of a given size category, the Census
gives information on the share of property tax in the composition
of revenues for 1962 and 1972. Although these shares have changed
between 1962 and 1972, extrapolation is difficult on the basis
of these changes. The municipalities which have already reduced
the share of the property tax through diversification are not likely
to continue diversifying. On the other hand, the municipalities
which have not broadened their tax base are much more likely to
diversify. Therefore, the 1972 shares are assumed to remain con-

147

stant. Property tax shares by population size are presented in Exhibit III-65.

The composition of the remainder of the revenues is assumed to be similar to that of the municipal total for each state. This information is summarized in Exhibit III-66. Public utilities and "other" taxes were excluded from the distributions used, because it did not seem likely that revenues from these sources would be used for financing water pollution control.

Financing O&M Charges (Matrix V)

On the basis of the questionnaire results, it was assumed that local governments finance 80% of the O&M costs of their sewerage facilities from sewer charges, and 20% from the property tax. It should be noted that although this assumption is not consistent with the requirements of the Act, which currently does not allow the use of ad valorem taxes for financing O&M costs, it is based on our survey results.

Financing Through Expenditure Reductions (Matrix V)

Questionnaire data, earlier econometric work, and budget analysis all suggest that local governments may well finance a portion of new water pollution abatement activity by reducing other expenditures. This expenditure substitution is, of course, relevant for the incidence patterns calculated.

In this research, we assumed that 10% of the local share of municipal costs would be financed through expenditure reduction. This is admittedly a rough estimate. Nevertheless, at least 10% of communities do admit some expenditure reduction, and more are likely to do so. Approximately 10% of the towns we surveyed in this project indicated that expenditure substitution was a likely financing tool. Furthermore, case studies, done on the Merrimack River Basin, found expenditure substitution in about 10% of the towns and cities in the sample.*

Expenditure reductions by special sewer districts are not relevant for distributional considerations, assuming that different categories of abatement projects benefit the same group. In other words, if a set of collector sewers was not built in order to finance an interceptor sewer mandated by P.L. 92-500, foregoing

---

*Oster & Roberts, "Determinants of the Distribution of Environmental Protection Expenditures among Local Governments: Water Pollution Control in the Merrimack Valley," paper presented at the NBER Conference on the Economic Analysis of Political Bahavior, April 12, 1975.

the benefits of the collector sewer is compensated by the benefits of the interceptor sewer.

For counties and municipal governments this is not the case, however. The survey results indicated that expenditure on police, fire, recreation, natural resources, and road maintenance may be reduced in order to pay for the necessary water pollution control expenditures. County and municipality expenditures on these services are described by Exhibits III-67 and III-68 respectively. We have assumed that 10% of their costs will be funded from reductions in these services, with the reductions being proportional to the relative magnitudes of these categories.

Incidence Assumptions (Matrix VI)

The distribution of the burden of the various federal, state, and local taxes depends on the underlying shifting assumptions. For the purposes of these calculations, we have selected Musgrave's benchmark shifting assumptions and the resulting distributional patterns. This has been discussed more fully in the "Incidence" section earlier in this chapter.

Adjusting the National Distribution of Burdens According to the Local Distribution of Income (Matrix VII)

Musgrave, Case and Leonard have calculated the distribution of tax burdens and expenditure benefits on the basis of the national distribution of income. Since the distribution is different at the local level, an adjustment is necessary. The Brown & Medoff formula* was used for this purpose. Data on the distribution of income was taken from the 1970 Census; Census income was transformed into total income through the use of the proportionality factors described in Exhibit II-3. The 15 income brackets used by the Census were reconciled with the 10 income brackets used in this study by assuming that the within-bracket distribution of income was linear.

With respect to changes over time, we have assumed that the distribution of income will remain constant at both the national and regional level. As income and population both increase over time, the bracket limits were "stretched" according to the Musgrave technique (described in Chapter 2), and the proportions are assumed to remain constant in each.

THE RESULTS

In the preceding section, a number of alternative assumptions were listed for some parts of the incidence calculations. For example, the costs of Needs categories I, II, and IVB, or categories I-V could

---

*The Brown & Medoff formula is described in the "Incidence" section of this chapter.

149

Exhibit III-65

Municipalities: Percentage of Total Tax Revenues From Property Tax — 1972

State	All Municipalities	Municipality Size Group (000s)				
		100	25-100	10-25	5-10	less than 5
Alabama	21.51 %	21.78%	21.62%	23.83%	18.83%	20.13 %
Alaska	56.72 %	0.00%	93.44%	42.99%	42.33%	47.93 %
Arizona	30.24 %	29.20%	30.28%	31.63%	38.90%	41.61 %
Arkansas	52.10 %	47.00%	53.72%	50.48%	56.02%	58.63 %
California	52.38 %	55.93%	45.64%	46.33%	56.16%	48.99 %
Colorado	38.45 %	41.99%	26.72%	35.72%	51.86%	46.02 %
Connecticut	99.13 %	99.10%	99.19%	99.12%	99.05%	98.95 %
Delaware	57.78 %	0.00%	50.84%	91.73%	94.73%	92.25 %
District of Columbia	30.88 %	30.88%	0.00%	0.00%	0.00%	0.00 %
Florida	52.22 %	56.52%	51.80%	45.65%	45.08%	47.88 %
Georgia	65.19 %	66.75%	64.73%	61.62%	60.57%	64.31 %
Hawaii	78.82 %	78.82%	0.00%	0.00%	0.0C%	0.00 %
Idaho	92.07 %	0.00%	93.40%	92.83%	88.00%	90.01 %
Illinois	56.00 %	61.54%	50.72%	47.28%	42.56%	47.78 %
Indiana	98.47 %	98.36%	99.43%	98.57%	98.47%	96.14 %
Iowa	94.64 %	93.48%	95.46%	95.77%	96.41%	93.21 %
Kansas	87.08 %	87.39%	82.32%	87.96%	85.31%	89.66 %
Kentucky	43.32 %	37.21%	40.07%	57.13%	45.56%	66.75 %
Louisiana	42.24 %	40.34%	46.86%	49.47%	39.92%	45.62 %
Maine	97.94 %	0.00%	98.86%	99.35%	99.48%	96.04 %
Maryland	75.80 %	73.86%	92.23%	96.66%	95.28%	80.83 %
Massachusetts	99.25 %	99.02%	99.45%	99.36%	99.04%	98.53 %
Michigan	75.70 %	60.89%	86.56%	97.04%	94.79%	96.51 %
Minnesota	87.45 %	84.39%	87.71%	90.97%	93.74%	92.57 %
Mississippi	86.99 %	92.69%	86.67%	85.85%	83.85%	80.59 %

Exhibit III-65 (Cont'd)

Municipalities: Percentage of Total Tax Revenues From Property Tax — 1972

State	All Municipalities	Municipality Size Group (000s)				
		100	25-100	10-25	5-10	less than 5
Missouri	35.10 %	29.84 %	43.04 %	37.95 %	51.24 %	61.61 %
Montana	91.62 %	0.00 %	92.27 %	91.04 %	91.02 %	91.18 %
Nebraska	73.83 %	64.03 %	91.64 %	93.19 %	87.18 %	91.22 %
Nevada	53.89 %	52.78 %	50.93 %	60.34 %	56.46 %	61.57 %
New Hampshire	96.63 %	0.00 %	98.45 %	97.91 %	96.94 %	92.45 %
New Jersey	80.96 %	85.92 %	83.08 %	75.82 %	71.93 %	72.05 %
New Mexico	69.47 %	78.41 %	59.07 %	46.46 %	51.13 %	58.14 %
New York	62.94 %	59.11 %	92.57 %	90.03 %	94.20 %	95.25 %
North Carolina	95.19 %	95.11 %	94.90 %	96.02 %	95.50 %	94.98 %
North Dakota	85.06 %	0.00 %	85.16 %	91.47 %	87.88 %	81.51 %
Ohio	37.73 %	33.38 %	39.48 %	41.65 %	44.27 %	56.17 %
Oklahoma	36.68 %	42.22 %	32.72 %	27.53 %	24.71 %	29.03 %
Oregon	81.88 %	78.49 %	86.46 %	82.40 %	84.30 %	84.70 %
Pennsylvania	46.56 %	35.32 %	67.05 %	62.07 %	61.13 %	61.86 %
Rhode Island	98.87 %	98.84 %	98.97 %	98.74 %	99.06 %	97.95 %
South Carolina	77.08 %	76.66 %	78.76 %	76.08 %	75.01 %	76.30 %
South Dakota	84.37 %	0.00 %	64.20 %	95.36 %	95.78 %	96.43 %
Tennessee	70.94 %	70.07 %	80.44 %	72.16 %	70.00 %	71.83 %
Texas	66.31 %	66.86 %	67.42 %	62.92 %	62.40 %	64.82 %
Utah	54.87 %	54.78 %	55.24 %	53.69 %	52.27 %	57.99 %
Vermont	97.35 %	0.00 %	94.20 %	97.40 %	97.72 %	97.75 %
Virginia	59.88 %	59.52 %	58.14 %	63.57 %	60.75 %	62.23 %
Washington	48.94 %	42.84 %	54.56 %	55.03 %	55.40 %	61.79 %
West Virginia	43.21 %	0.00 %	43.62 %	38.16 %	41.92 %	47.74 %
Wisconsin	97.70 %	97.23 %	98.96 %	99.04 %	98.14 %	93.64 %
Wyoming	71.09 %	0.00 %	68.98 %	67.43 %	74.46 %	75.36 %

Exhibit III-66

Municipalities: Percent Distribution of Non-Property Taxes, 1972

State	General Sales	Selective Sales	Public Utilities	Motor Vehicle Licenses	Income	Other	Total
Alabama	44.77 %	12.83 %	3.87 %	2.75 %	7.75 %	28.02 %	100%
Alaska	78.39 %	0 %	7.84 %	0.05 %	0 %	13.71 %	100%
Arizona	71.49 %	10.48 %	7.95 %	0 %	0 %	10.07 %	100%
Arkansas	3.90 %	5.73 %	50.63 %	0.30 %	0 %	39.44 %	100%
California	56.00 %	4.62 %	16.18 %	0 %	0 %	23.20 %	100%
Colorado	68.86 %	4.17 %	7.93 %	0.28 %	0 %	18.75 %	100%
Connecticut	0 %	0 %	0 %	0 %	0 %	100.00 %	100%
Delaware	25.92 %	0 %	0.87 %	0 %	72.40 %	26.74 %	100%
D.C.	0 %	18.34 %	3.77 %	4.14 %	43.29 %	4.55 %	100%
Florida	0 %	35.27 %	44.17 %	0.59 %	0.04 %	19.93 %	100%
Georgia	0 %	37.17 %	15.38 %	0.95 %	0 %	46.49 %	100%
Hawaii	0 %	33.35 %	9.61 %	46.22 %	0 %	10.83 %	100%
Idaho	0 %	0 %	33.29 %	0 %	0 %	66.71 %	100%
Illinois	55.36 %	1.18 %	21.49 %	11.04 %	0 %	10.94 %	100%
Indiana	0 %	0 %	0 %	23.27 %	0 %	76.73 %	100%
Iowa	0 %	0 %	18.01 %	0 %	0.07 %	81.91 %	100%
Kansas	6.05 %	0 %	59.84 %	0 %	0 %	34.11 %	100%
Kentucky	0 %	3.79 %	1.51 %	4.71 %	75.59 %	14.39 %	100%
Louisiana	68.22 %	5.30 %	7.15 %	1.05 %	0 %	18.28 %	100%
Maine	0 %	0 %	0 %	0 %	0 %	100.00 %	100%
Maryland	0 %	5.45 %	22.37 %	0 %	51.89 %	20.29 %	100%
Massachusetts	0 %	0 %	0 %	0.52 %	0 %	99.48 %	100%
Michigan	0 %	0 %	10.04 %	0.09 %	81.13 %	8.74 %	100%
Minnesota	6.86 %	9.08 %	37.78 %	0.56 %	0 %	45.72 %	100%
Mississippi	0 %	0 %	37.66 %	2.15 %	0 %	60.19 %	100%

152

Exhibit III-66 (Cont'd)

Municipalities: Percent Distribution of Non-Property Taxes, 1972

State	General Sales	Selective Sales	Public Utilities	Motor Vehicle Licenses	Income	Other	Total
Missouri	16.77%	5.62 %	27.71 %	3.64 %	32.47 %	13.80 %	100%
Montana	0%	0 %	0 %	34.20 %	0 %	65.80 %	100%
Nebraska	57.70%	0 %	11.90 %	12.09 %	0 %	18.30 %	100%
Nevada	0%	4.28 %	21.86 %	0	0 %	73.86 %	100%
New Hampshire	0%	0 %	0 %	0	0 %	100.00 %	100%
New Jersey	0%	8.03 %	72.15 %	0.02 %	0 %	19.79 %	100%
New Mexico	0%	17.39 %	30.38 %	0 %	0 %	52.23 %	100%
New York	31.37%	13.94 %	3.91 %	0.29 %	45.96 %	4.55 %	100%
North Carolina	0%	0 %	0.36 %	9.15 %	0 %	90.48 %	100%
North Dakota	0%	0 %	13.14 %	0	0 %	86.86 %	100%
Ohio	0%	0.93 %	0.12 %	0.19 %	89.77 %	8.99 %	100%
Oklahoma	76.40%	0.25 %	12.30 %	0	0 %	11.05 %	100%
Oregon	0%	1.96 %	48.13 %	0	0 %	49.91 %	100%
Pennsylvania	0%	1.96 %	0.05 %	0.01 %	78.05 %	19.94 %	100%
Rhode Island	0%	0 %	0 %	0	0 %	100.00 %	100%
South Carolina	0%	0 %	0.29 %	0	0 %	99.71 %	100%
South Dakota	76.42%	0 %	0 %	0.34 %	0 %	23.23 %	100%
Tennessee	36.55%	27.29 %	6.40 %	15.50 %	0 %	14.25 %	100%
Texas	69.80%	1.53 %	20.28 %	0	0 %	8.38 %	100%
Utah	60.84%	0 %	17.74 %	3.42 %	0 %	18.01 %	100%
Vermont	0%	0 %	0 %	0	0 %	100.00 %	100%
Virginia	28.54%	7.89 %	32.77 %	5.31 %	0 %	25.48 %	100%
Washington	33.32%	3.50 %	32.93 %	0.15 %	0 %	30.09 %	100%
West Virginia	0%	5.08 %	8.16 %	0.09 %	0 %	86.66 %	100%
Wisconsin	0%	9.09 %	0 %	0.14 %	0 %	90.77 %	100%
Wyoming	0%	0	34.26 %	0	0 %	65.74 %	100%

Exhibit III-67

Expenditures of County Governments by Function, By State, 1971-72

REGION State	Total ($000)	Police, Fire, Parks and Recreation, Natural Resources ($000)	% of Total	Highway, Non-Capital ($000)	% of Total
United States	$23,931,899	$1,847,046	7.7%	$1,811,442	7.6%
NORTHEAST	4,396,964	288,445	6.6%	185,921	4.2%
New England					
Maine	8,213	1,346	16.4%	48,934	17.1%
New Hampshire	16,981	1,219	7.2%	—	0.0%
Vermont	398	7	1.8%	—	0.0%
Massachusetts	95,431	2,405	2.5%	4,946	5.2%
Connecticut					
Rhode Island					
Middle Atlantic					
New York	2,755,486	216,480	7.9%	118,421	4.3%
New Jersey	1,031,154	42,076	4.1%	41,124	4.0%
Pennsylvania	489,310	24,915	5.1%	20,804	4.3%
Delaware	37,182	8,597	23.1%	—	0.0%
NORTH CENTRAL	5,548,002	360,688	6.5%	739,687	6.9%
East North Central					
Ohio	862,669	36,602	4.2%	108,323	12.6%
Indiana	452,042	21,461	4.8%	45,895	10.2%
Illinois	714,182	50,402	7.1%	64,752	9.1%
Wisconsin	730,535	55,836	7.6%	132,750	18.2%
Michigan	996,941	84,638	8.5%	101,316	10.2%
West North Central					
Minnesota	689,141	28,576	4.2%	90,909	13.2%
Iowa	319,616	16,298	5.1%	68,057	21.3%
Missouri	224,045	38,982	17.4%	28,145	12.6%
North Dakota	59,620	3,551	6.0%	16,334	27.4%
South Dakota	49,202	3,219	6.5%	16,660	33.9%
Nebraska	138,210	8,552	6.2%	31,705	22.9%
Kansas	311,800	12,568	4.0%	34,739	11.1%

154

Exhibit III-67 (Cont'd)

Expenditures of County Governments by Function, By State, 1971-72

REGION State	Total ($000)	Police, Fire, Parks and Recreation, Natural Resources ($000)	% of Total	Highway, Non-Capital ($000)	% of Total
SOUTH					
South Atlantic					
Maryland	$ 1,443,542	$ 92,937	6.4 %	$ 37,746	2.6 %
West Virginia	51,704	6,741	13.0 %	14	0.0 %
Virginia	915,735	48,909	5.3 %	6,246	6.8 %
North Carolina	1,388,520	27,941	2.0 %	--	0.0 %
South Carolina	179,577	14,195	7.9 %	9,792	5.5 %
Georgia	311,871	37,531	12.0 %	40,711	13.1 %
Florida	633,504	93,828	14.8 %	42,801	6.8 %
East South Central					
Kentucky	104,768	46,263	10.3 %	79,125	17.6 %
Tennessee	690,882	18,576	2.7 %	47,949	6.9 %
Mississippi	207,858	4,087	2.0 %	46,020	22.1 %
Alabama	186,573	15,892	8.5 %	43,734	23.4 %
West South Central					
Arkansas	91,084	5,298	5.8 %	21,512	23.6 %
Louisiana	286,589	44,323	15.5 %	48,934	17.1 %
Oklahoma	143,090	7,812	5.5 %	54,182	37.9 %
Texas	449,236	46,263	10.3 %	79,125	17.6 %
WEST					
Mountain					
Montana	126,362	7,943	6.3 %	12,291	1.0 %
Wyoming	43,415	3,011	6.9 %	3,637	8.4 %
Colorado	200,612	10,985	5.5 %	30,557	15.2 %
New Mexico	47,864	4,384	9.2 %	6,058	12.7 %
Idaho	70,599	5,813	8.2 %	14,399	20.4 %
Nevada	128,840	25,305	19.6 %	5,088	4.0 %
Utah	56,225	14,117	25.1 %	7,144	12.7 %
Arizona	186,817	16,188	8.7 %	14,857	8.0 %
Pacific					
Washington	293,154	46,471	15.9 %	68,172	23.3 %
Oregon	196,450	21,230	10.8 %	42,240	21.5 %
California	5,324,234	532,629	10.0 %	186,226	3.5 %
Alaska	141,207	3,017	2.1 %	1,345	1.0 %
Hawaii	49,438	21,489	43.5 %	2,936	5.9 %

Exhibit III-68

Expenditure Reductions for Financing Treatment Requirements By Community Size

REGION State	All Muni- cipalities	100,000 and over	25,000- 99,999	10,000- 24,999	5,000- 9,999	less than 5,000
UNITED STATES						
Total Expenditures	$35,899,853	$24,112,102	$6,368,068	$2,540,982	$1,218,714	$1,659,999
Expenditures--Selected Services*	9,105,866	5,620,076	1,922,288	686,709	370,292	369,837
Expenditures--Highways**	1,614,738	631,089	394,558	237,776	141,303	220,007
NORTHEAST						
New England						
Maine						
Total Expenditures	126,070	--	57,418	50,809	14,371	3,422
Expenditures--Selected Services	18,007	--	9,144	6,652	1,826	385
Expenditures--Highways	9,334	--	3,005	4,648	1,304	376
New Hampshire						
Total Expenditures	101,061	--	65,402	31,123	5,536	--
Expenditures--Selected Services	18,335	--	12,732	4,607	998	--
Expenditures--Highways	7,413	--	4,416	2,339	658	--
Vermont						
Total Expenditures	21,959	--	5,620	5,220	5,526	5,593
Expenditures--Selected Services	6,093	--	2,393	1,766	868	1,067
Expenditures--Highways	4,043	--	553	1,173	998	1,319
Massachusetts						
Total Expenditures	1,498,131	772,958	702,086	23,087	--	--
Expenditures--Selected Services	242,773	155,256	130,035	2,896	--	--
Expenditures--Highways	48,602	16,021	31,247	1,333	--	--

*These are comprised of police and fire protection, sanitation other than sewerage, parks and recreation.

**Expenditures on highways exclude capital outlays.

Exhibit III-68

Expenditure Reductions for Financing Treatment Requirements By Community Size (Cont'd)

REGION State	All Muni- cipalities	100,000 and over	25,000- 99,999	10,000- 24,999	5,000- 9,999	less than 5,000
Connecticut						
Total Expenditures	$ 667,617	$ 337,116	$260,236	$ 20,820	$ 3,986	$ 2,204
Expenditures--Selected Services	128,135	84,313	38,998	3,119	873	524
Expenditures--Highways	17,872	7,524	7,904	1,231	539	229
Rhode Island						
Total Expenditures	200,870	73,340	123,465	4,065	--	--
Expenditures--Selected Services	42,834	19,878	22,144	811	--	--
Expenditures--Highways	6,864	2,612	4,116	136	--	--
Middle Atlantic						
New York						
Total Expenditures	10,438,165	9,749,719	380,395	150,558	75,495	81,999
Expenditures--Selected Services	1,566,526	1,372,836	102,975	47,634	22,840	20,240
Expenditures--Highways	153,706	99,279	18,337	13,984	8,857	13,248
New Jersey						
Total Expenditures	1,294,380	534,187	423,671	168,041	99,633	68,217
Expenditures--Selected Services	336,654	126,187	109,604	56,452	28,705	15,708
Expenditures--Highways	48,542	7,087	11,504	15,099	9,228	5,623
Pennsylvania						
Total Expenditures	1,218,653	829,193	128,122	71,397	65,694	61,067
Expenditures--Selected Services	410,049	294,216	45,259	29,277	23,938	17,358
Expenditures--Highways	81,641	34,179	11,713	11,533	11,695	12,522
Delaware						
Total Expenditures	66,489	--	55,173	5,307	1,570	4,439
Expenditures--Selected Services	15,931	--	11,603	2,291	581	1,407
Expenditures--Highways	2,905	--	1,662	354	246	643

*These are comprised of police and fire protection, sanitation other than sewerage, parks and recreation.

**Expenditures on highways exclude capital outlays.

Exhibit III-68

Page 3 of 9

Expenditure Reductions for Financing Treatment Requirements By Community Size (Cont'd)

REGION State	All Muni- cipalities	100,000 and over	25,000- 99,999	10,000- 24,999	5,000- 9,999	less than 5,000
NORTH CENTRAL						
East North Central						
Ohio						
Total Expenditures	$1,336,409	$752,883	$268,580	$175,092	$63,608	$76,246
Expenditures--Selected Services	450,516	255,589	88,025	62,554	23679	20,669
Expenditures--Highways	93,551	36,006	18,748	20,354	7,892	10,551
Indiana						
Total Expenditures	536,820	300,544	110,970	59,966	27,644	37,694
Expenditures--Selected Services	170,053	91,257	39,180	20,511	9,754	9,351
Expenditures--Highways	42,584	20,531	4,528	6,332	3,122	5,072
Illinois						
Total Expenditures	1,483,103	891,261	294,703	150,270	60,824	66,044
Expenditures--Selected Services	578,421	407,107	114,094	56,589	20,940	16,614
Expenditures--Highways	126,018	46,988	33,757	21,069	10,042	14,163
Wisconsin						
Total Expenditures	918,838	256,058	354,038	155,251	66,690	86,801
Expenditures--Selected Services	201,787	81,941	66,434	24,726	13,801	18,759
Expenditures--Highways	53,875	10,490	16,248	8,472	6,337	12,328
Michigan						
Total Expenditures	1,363,144	725,295	27,919	14,754	8,411	27,819
Expenditures--Selected Services	426,888	235,618	118,719	41,773	14,511	17,265
Expenditures--Highways	85,193	25,403	27,099	13,063	7,830	11,798
West North Central						
Minnesota						
Total Expenditures	557,121	214,646	115,621	86,159	46,539	94,156
Expenditures--Selected Services	134,733	62,951	30,655	19,810	8,923	12,396
Expenditures--Highways	42,275	13,329	8,854	6,354	4,127	9,611

Exhibit III-68

Page 4 of 9

Expenditure Reductions for Financing Treatment Requirements By Community Size (Cont'd)

REGION State	All Municipalities	100,000 and over	25,000-99,999	10,000-24,999	5,000-9,999	less than 5,000
Iowa						
Total Expenditures	$296,770	$ 61,415	$112,686	$ 27,480	$31,306	$ 63,883
Expenditures--Selected Services	75,530	19,729	31,769	6,442	8,029	9,563
Expenditures--Highways	34,731	7,193	10,637	2,671	4,108	10,079
Missouri						
Total Expenditures	634,246	376,194	50,376	49,046	47,582	111,049
Expenditures--Selected Services	179,217	119,073	20,812	18,898	10,625	9,810
Expenditures--Highways	35,941	14,233	4,924	6,199	4,164	6,403
North Dakota						
Total Expenditures	49,644	--	29,619	5,288	2,797	11,940
Expenditures--Selected Services	13,760	--	7,914	2,893	834	2,120
Expenditures--Highways	4,087	--	1,695	484	346	1,562
South Dakota						
Total Expenditures	49,882	--	20,808	9,645	3,629	15,800
Expenditures--Selected Services	13,575	--	7,437	2,715	729	2,693
Expenditures--Highways	6,449	--	1,976	1,384	454	2,634
Nebraska						
Total Expenditures	165,795	92,528	5,366	18,107	11,570	38,224
Expenditures--Selected Services	46,562	30,321	1,128	6,153	3,755	5,203
Expenditures--Highways	15,157	5,313	355	2,194	1,806	5,489
SOUTH						
South Atlantic						
Maryland						
Total Expenditures	803,199	724,295	27,919	14,754	8,411	27,819
Expenditures--Selected Services	138,593	116,218	10,295	6,393	3,018	2,671
Expenditures--Highways	20,976	13,053	2,694	1,808	1,464	1,956

Exhibit III-68

Expenditure Reductions for Financing Treatment Requirements By Community Size (Cont'd)

REGION State	All Municipalities	100,000 and over	25,000-99,999	10,000-24,999	5,000-9,999	less than 5,000
West Virginia						
Total Expenditures	$ 93,773	--	$ 59,644	$10,805	$ 9,270	$14,053
Expenditures--Social Services	23,381	--	17,748	4,573	2,154	3,907
Expenditures--Highways	8,537	--	4,096	1,448	849	2,143
Virginia						
Total Expenditures	928,139	$620,864	157,894	89,216	35,947	24,219
Expenditures--Social Services	132,893	85,360	23,189	12,605	6,639	5,100
Expenditures--Highways	26,362	12,548	5,329	3,546	2,452	2,486
North Carolina						
Total Expenditures	375,426	162,111	93,068	45,892	26,146	48,209
Expenditures--Social Services	134,163	55,959	35,602	21,119	9,272	12,208
Expenditures--Highways	28,820	7,974	6,908	5,155	2,740	6,043
South Carolina						
Total Expenditures	103,433	22,163	34,232	13,406	18,517	15,117
Expenditures--Social Services	48,373	7,530	16,427	7,356	9,601	6,823
Expenditures--Highways	6,711	698	2,206	1,129	1,642	1,032
Georgia						
Total Expenditures	398,736	245,225	41,007	47,942	17,760	46,802
Expenditures--Social Services	145,640	81,190	18,452	21,271	8,425	16,301
Expenditures--Highways	24,159	9,201	3,813	3,993	2,259	4,894
Florida						
Total Expenditures	712,399	308,427	228,287	86,045	44,721	44,948
Expenditures--Social Services	292,749	125,391	91,909	40,002	20,932	14,514
Expenditures--Highways	41,876	17,121	10,270	5,590	4,799	4,097
East South Central						
Kentucky						
Total Expenditures	220,222	103,044	48,987	29,995	23,096	21,110
Expenditures--Social Services	72,010	38,867	12,834	9,148	5,215	5,946
Expenditures--Highways	10,148	2,791	2,341	1,878	1,437	1,729

Exhibit III-68

Page 6 of 9

Expenditure Reductions for Financing Treatment Requirements By Community Size (Cont'd)

REGION State	All Muni- cipalities	100,000 and over	25,000- 99,999	10,000- 24,999	5,000- 9,999	less than 5,000
Tennessee						
Total Expenditures	$740,421	546,111	59,750	64,739	35,482	34,338
Expenditures--Social Services	149,763	107,526	13,976	12,909	7,700	7,654
Expenditures--Highways	27,659	14,064	2,169	3,860	3,536	4,030
Mississippi						
Total Expenditures	139,565	25,724	43,264	29,688	14,025	26,865
Expenditures--Social Services	40,527	10,439	12,507	8,562	3,678	5,342
Expenditures--Highways	13,023	1,455	3,063	2,980	2,187	3,338
Alabama						
Total Expenditures	349,023	148,805	87,614	39,719	43,111	29,775
Expenditures--Social Services	98,503	48,744	20,632	10,899	10,019	7,412
Expenditures--Highways	23,678	7,055	4,827	3,479	4,138	4,178
West South Central						
Arkansas						
Total Expenditures	112,923	16,740	38,975	22,406	16,214	8,589
Expenditures--Social Services	32,645	7,465	12,131	5,852	3,096	4,099
Expenditures--Highways	12,776	1,282	3,342	2,679	1,763	3,690
Louisiana						
Total Expenditures	299,105	192,664	47,630	25,699	14,206	18,908
Expenditures--Social Services	115,118	73,182	22,111	10,357	5,113	4,355
Expenditures--Highways	20,051	10,363	2,145	3,336	1,823	2,384
Oklahoma						
Total Expenditures	274,329	123,696	52,014	48,887	21,670	28,063
Expenditures--Social Services	81,129	41,712	15,646	10,157	6,434	7,180
Expenditures--Highways	16,293	5,661	2,164	2,490	2,491	3,487

Exhibit III-68

Expenditure Reductions for Financing Treatment Requirements by Community Size (Cont'd)

REGION State	All Municipalities	100,000 and over	25,000-99,999	10,000-24,999	5,000-9,999	less than 5,000
Texas						
Total Expenditures	$1,164,837	$674,059	$272,539	$104,373	$47,847	$66,019
Expenditures--Social Services	454,853	265,160	112,991	39,713	19,614	17,374
Expenditures--Highways	66,636	27,730	15,477	10,157	5,912	7,350
WEST						
Mountain						
Montana						
Total Expenditures	45,444	--	21,721	10,107	2,990	10,725
Expenditures--Social Services	15,733	--	7,654	3,809	1,466	2,803
Expenditures--Highways	5,357	--	2,426	929	498	1,504
Wyoming						
Total Expenditures	24,086	--	8,377	3,838	3,779	8,092
Expenditures--Social Services	9,074	--	3,781	1,738	1,540	2,015
Expenditures--Highways	3,039	--	820	433	731	1,054
Colorado						
Total Expenditures	352,489	245,276	55,826	20,556	8,433	22,397
Expenditures--Social Services	89,340	52,210	21,646	5,883	3,385	5,214
Expenditures--Highways	18,135	6,420	5,657	1,742	1,064	3,253
New Mexico						
Total Expenditures	116,991	65,170	24,204	13,263	5,091	9,264
Expenditures--Social Services	45,453	26,001	8,722	5,551	2,426	2,754
Expenditures--Highways	6,997	2,238	1,771	1,255	735	908

Exhibit III-68

Expenditure Reductions for Financing Treatment Requirements by Community Size (Cont'd)

REGION State	All Municipalities	100,000 and over	25,000- 99,999	10,000- 24,999	5,000- 9,999	less than 5,000
Idaho						
Total Expenditures	$ 44,304	--	$ 20,655	$ 8,087	$ 2,679	$12,883
Expenditures--Social Services	14,989	--	9,071	3,256	1,172	3,491
Expenditures--Highways	5,426	--	2,259	792	277	2,097
Nevada						
Total Expenditures	70,494	$ 29,531	23,027	11,638	2,597	3,700
Expenditures--Social Services	30,538	14,584	10,166	3,708	1,025	1,055
Expenditures--Highways	5,161	1,311	2,342	796	313	417
Utah						
Total Expenditures	69,277	25,745	15,622	7,950	8,031	11,929
Expenditures--Social Services	26,092	10,152	6,698	3,818	2,836	2,588
Expenditures--Highways	7,324	2,437	1,156	1,047	1,202	1,484
Arizona						
Total Expenditures	247,015	174,924	45,003	7,334	11,227	8,527
Expenditures--Social Services	111,820	82,892	19,139	3,076	4,261	2,451
Expenditures--Highways	13,146	6,970	2,455	701	1,651	1,369
Pacific						
Washington						
Total Expenditures	393,179	215,970	67,061	61,364	13,817	34,967
Expenditures--Social Services	158,077	93,992	26,598	21,800	5,649	10,027
Expenditures--Highways	27,803	11,354	5,737	4,510	1,521	4,680

Exhibit III-68

Expenditure Reductions for Financing Treatment Requirements by Community Size (Cont'd)

REGION State	All Municipalities	100,000 and over	25,000-99,999	10,000-24,999	5,000-9,999	less than 5,000
Oregon						
Total Expenditures	$ 193,139	$ 74,789	$ 44,056	$ 32,731	$16,201	$ 25,363
Expenditures--Social Services	73,440	33,417	17,236	11,768	4,918	6,101
Expenditures--Highways	15,639	4,403	3,922	2,822	1,770	2,723
California						
Total Expenditures	2,987,530	1,912,124	729,053	242,042	52,750	51,561
Expenditures--Social Services	1,157,765	698,829	322,523	101,088	19,801	15,524
Expenditures--Highways	185,640	83,333	64,695	24,631	5,880	7,371
Alaska						
Total Expenditures	349,023	148,805	87,614	39,719	43,111	29,775
Expenditures--Social Services	98,503	48,744	20,632	10,899	10,019	7,412
Expenditures--Highways	23,678	7,055	4,827	3,479	4,138	4,178
Hawaii						
Total Expenditures	114,428	148,428	--	--	--	--
Expenditures--Social Services	55,914	55,914	--	--	--	--
Expenditures--Highways	8,258	8,258	--	--	--	--
D.C.						
Total Expenditures	1,115,064	1,115,064	--	--	--	--
Expenditures--Social Services	145,713	145,713	--	--	--	--
Expenditures--Highways	21,266	21,266	--	--	--	--

be considered; either a zero or the Commission's high baseline could be included in the calculations, and so on. Given these alternatives, we have calculated the distribution of the cost burden for five different scenarios. These are:

(1) Needs Survey categories I, II, and IVB, with zero baselines, for 1977, 1980, and 1985.

(2) Needs Survey categories I-V, with zero baselines to 1985.

(3) Needs Survey categories, I, II, and IVB, with high baselines included, for 1985.

(4) Needs Survey categories I, II, and IVB, with zero baselines for 1985, for counties, special districts and municipalities.

(5) Needs Survey categories I, II, and IVB, with zero baselines by five city sizes for 1985.

Before turning to discuss the results for each of these scenarios, it should be noted that, a priori, we should not expect a smooth progression or distribution of burdens. The tables in the "Incidence" section of this chapter described the distributions of the tax burdens and expenditure benefits. These distributions are uneven across income groups. The distribution of the tax burden for water pollution control, which is a composite of taxes and expenditure cuts, may also be expected to be unevenly distributed across income groups.

It should also be noted that all cost figures are presented in 1973 constant dollars.

Needs Survey Categories I, II, and IVB, with Zero Baselines, for 1977, 1980, and 1985

The distribution of the costs for the 10 income groups, in the four geographical regions and the U.S. is presented in Exhibit III-69 for 1977, Exhibit III-70 for 1980, and in Exhibit III-71 for 1985.

Exhibit III-69 shows that in 1977 families in the lowest income group will pay $133.2 million toward water pollution control. On a per family basis, this implies $16.55 which in turn is 0.51% of the mean (total) income in that income group. The per family tax burden for the median (fifth) income group is $51.10 or 0.28% of their (total) income. The per family tax burdens necessary for financing Categories I, II, and IVB of the Needs Survey are calculated for each income group for 1977, 1980, and 1985. These figures are presented in Exhibit III-72.

Gini coefficients have been calculated for 1977, 1980, and 1985. In these calculations, the distribution of income in the

Exhibit III-69

1977 Distribution of the Costs of Publicly-Owned Wastewater Treatment Facilities by Income Group, for the U.S. and Census Regions: Categories I, II, and IVB -- Zero Baselines

REGION	INCOME GROUP									
	1	2	3	4	5	6	7	8	9	10
UNITED STATES										
Total ($Millions)*	$133.2	$131.8	$236.7	$370.4	$474.7	$1075.9	$559.9	$390.3	$383.0	$510.5
Percent of Total**	3.1%	3.1%	5.5%	8.7%	11.1%	25.2%	13.1%	9.1%	9.0%	12.0%
NORTHEAST										
Total ($Millions)	$ 20.7	$ 20.7	$ 46.8	$ 80.6	$141.8	$ 273.6	$155.3	$109.1	$114.4	$159.0
Percent of Total	1.9%	1.9%	4.3%	7.4%	10.2%	25.0%	14.2%	10.0%	10.5%	14.5%
NORTH CENTRAL										
Total ($Millions)	$ 26.2	$ 26.7	$ 49.6	$ 85.1	$120.4	$ 286.1	$145.2	$100.8	$ 89.2	$116.5
Percent of Total	2.5%	2.6%	4.7%	8.1%	11.5%	27.4%	13.9%	9.6%	8.5%	11.1%
SOUTH										
Total ($Millions)	$ 70.1	$ 64.3	$108.3	$152.9	$173.8	$ 344.7	$158.2	$109.9	$108.4	$151.3
Percent of Total	4.9%	4.5%	7.5%	10.6%	12.1%	23.9%	11.0%	7.6%	7.5%	10.5%
WEST										
Total ($Millions)	$ 14.8	$ 16.1	$ 28.9	$ 47.2	$ 63.1	$ 155.1	$ 87.0	$ 60.5	$ 59.4	$ 74.3
Percent of Total	2.4%	2.7%	4.8%	7.8%	10.4%	25.6%	14.3%	10.0%	9.8%	12.2%
ALA., HAW., D.C.										
Total ($Millions)	$ 1.4	$ 1.4	$ 3.1	$ 4.6	$ 5.7	$ 16.4	$ 13.9	$ 10.0	$ 11.6	$ 9.4
Percent of Total	1.8%	1.8%	4.0%	6.0%	7.3%	21.1%	18.0%	12.9%	15.0%	12.1%

*Costs in millions of $1973

**Percent of each region's total expenditures borne by each income group.

Exhibit III-70

1980 Distribution of the Costs of Publicly-Owned Wastewater Treatment Facilities by Income Group, for the U.S. and Census Regions: Categories I, II, and IVB -- Zero Baselines

REGION	INCOME GROUP									
	1	2	3	4	5	6	7	8	9	10
UNITED STATES										
Total ($Millions)*	$182.4	$165.2	$285.4	$437.6	$577.1	$1236.3	$630.0	$448.6	$417.7	$539.9
Percent of Total**	3.7%	3.4%	5.8%	8.9%	11.4%	25.2%	12.9%	9.2%	8.5%	11.0%
NORTHEAST										
Total ($Millions)	$ 28.2	$ 29.0	$ 56.2	$ 95.1	$131.3	$ 315.0	$175.3	$125.7	$125.4	$169.0
Percent of Total	2.3%	2.3%	4.5%	7.6%	10.5%	25.2%	14.0%	10.1%	10.0%	13.6%
NORTH CENTRAL										
Total ($Millions)	$ 36.1	$ 33.6	$ 60.0	$100.8	$141.7	$ 329.8	$163.8	$116.5	$ 97.6	$123.6
Percent of Total	3.0%	2.8%	5.0%	8.4%	11.8%	27.4%	13.6%	9.7%	8.1%	10.3%
SOUTH										
Total ($Millions)	$ 95.8	$ 80.5	$130.2	$180.0	$202.9	$ 398.2	$176.6	$125.0	$117.0	$158.1
Percent of Total	5.8%	4.9%	7.8%	10.8%	12.2%	23.7%	10.6%	7.5%	7.0%	9.5%
WEST										
Total ($Millions)	$ 20.4	$ 20.3	$ 35.1	$ 56.2	$ 74.5	$ 179.3	$ 98.3	$ 69.8	$ 64.9	$ 78.6
Percent of Total	2.9%	2.9%	5.0%	8.1%	10.7%	25.7%	14.1%	10.0%	9.3%	11.3%
ALA., HAW., D.C.										
Total ($Millions)	$ 1.9	$ 1.8	$ 3.8	$ 5.6	$ 6.8	$ 19.1	$ 15.9	$ 11.6	$ 12.8	$ 10.0
Percent of Total	2.1%	2.0%	4.3%	6.2%	7.6%	21.4%	17.8%	13.0%	14.4%	11.3%

*Costs in millions of $1973.
**Percent of each region's total expenditures borne by each income group.

Exhibit III-71

1985 Distribution of the Costs of Publicly-Owned Wastewater Treatment Facilities by Income Group, for the U.S. and Census Regions: Categories I, II, and IVB -- Zero Baselines

REGION	INCOME GROUP									
	1	2	3	4	5	6	7	8	9	10
UNITED STATES										
Total ($Millions)*	$231.6	$198.9	$334.3	$505.1	$640.1	$1397.7	$701.1	$506.8	$451.2	$567.5
Percent of Total**	4.2%	3.6%	6.0%	9.1%	11.6%	25.3%	12.7%	9.2%	8.2%	10.3%
NORTHEAST										
Total ($Millions)	$ 35.8	$ 34.8	$ 65.8	$109.8	$150.9	$ 356.7	$195.6	$142.3	$136.1	$179.4
Percent of Total	2.5%	2.5%	4.7%	7.8%	10.7%	25.3%	13.9%	10.1%	9.7%	12.7%
NORTH CENTRAL										
Total ($Millions)	$ 46.1	$ 40.6	$ 70.5	$116.6	$163.1	$ 373.7	$182.6	$132.1	$105.6	$130.2
Percent of Total	3.4%	3.0%	5.2%	8.6%	12.0%	27.5%	13.4%	9.7%	7.8%	9.6%
SOUTH										
Total ($Millions)	$121.3	$ 96.8	$152.2	$207.1	$232.2	$ 442.0	$195.1	$140.1	$125.3	$164.5
Percent of Total	6.5%	5.2%	8.1%	11.0%	12.4%	23.6%	10.4%	7.5%	6.7%	8.8%
WEST										
Total ($Millions)	$ 26.1	$ 24.6	$ 41.3	$ 65.2	$ 86.0	$ 203.6	$109.8	$ 79.1	$ 70.2	$ 82.7
Percent of Total	3.3%	3.1%	5.2%	8.3%	10.9%	25.8%	13.9%	10.0%	8.9%	10.5%
ALA., HAW., D.C.										
Total ($Millions)	$ 22.4	$ 2.2	$ 4.5	$ 6.5	$ 7.8	$ 21.8	$ 17.8	$ 13.3	$ 14.0	$ 10.6
Percent of Total	2.4%	2.1%	4.5%	6.4%	7.8%	21.6%	17.7%	13.2%	13.9%	10.6%

*Costs in millions of $1973.
**Percent of each region's total expenditures paid by each income group.

```
                        Exhibit III-72

                   Annual Costs Per Family for

                   Needs Categories I, II, IVB

                    (in 1973 constant dollars)

Income Group            1977          1980          1985

      1              $  16.55      $  22.16      $  26.77
      2                 28.97         35.22         40.32
      3                 37.22         43.57         48.52
      4                 47.55         54.56         59.92
      5                 51.10         58.21         63.63
      6                 67.62         75.43         81.12
      7                110.57        120.92        127.94
      8                 70.71         78.84         87.75
      9                196.41        207.81        213.84
     10                981.73        999.81       1013.39
```

base case was adjusted for the distribution of the tax burdens
calculated in Exhibits III-69 through III-71. In each case, the
change in the Gini coefficient was less than .001; in other words,
it was not significant. Since the annual water pollution control
expenditures account for a very small fraction of total personal
income, this outcome appears reasonable.

The percentage distribution of the tax burdens are also shown
in Exhibits III-69 and III-71. In 1977, for the U.S., 6.2% of the
burden was borne by the lowest two income groups. This may be
compared with the median (fifth) income group which is responsible
for 11.1% of the burden, or the highest income group which pays
12.0% of the costs. These figures, however, may be misleading;
the fraction of families paying a given fraction of costs in the
lowest two income groups, indicate a relatively low per family cost.
For the median income group, 14.3% of families pay 11.1% of the cost,
and 0.8% of families in the highest income group pay 12.0% of the
costs.

Exhibit III-73 presents a Lorenz Curve which compares the distribution of the burden for 1977 with the distribution of the population. This may be compared with the distribution of the federal
tax burden. The distribution of the burden of pollution control
expenditures is roughly comparable to the distribution of the total
federal tax burden.

In order to examine trends in the distribution of burdens over
time, Exhibit III-70 for 1980, and Exhibit III-71 for 1985, may be compared with Exhibit III-69 for 1977. A noteworthy feature of this
comparison is that the overall distribution of burdens becomes more

EXHIBIT III-73

Lorenz Curves: The Distribution of Pollution Control Costs
Compared with the Federal Tax Burden

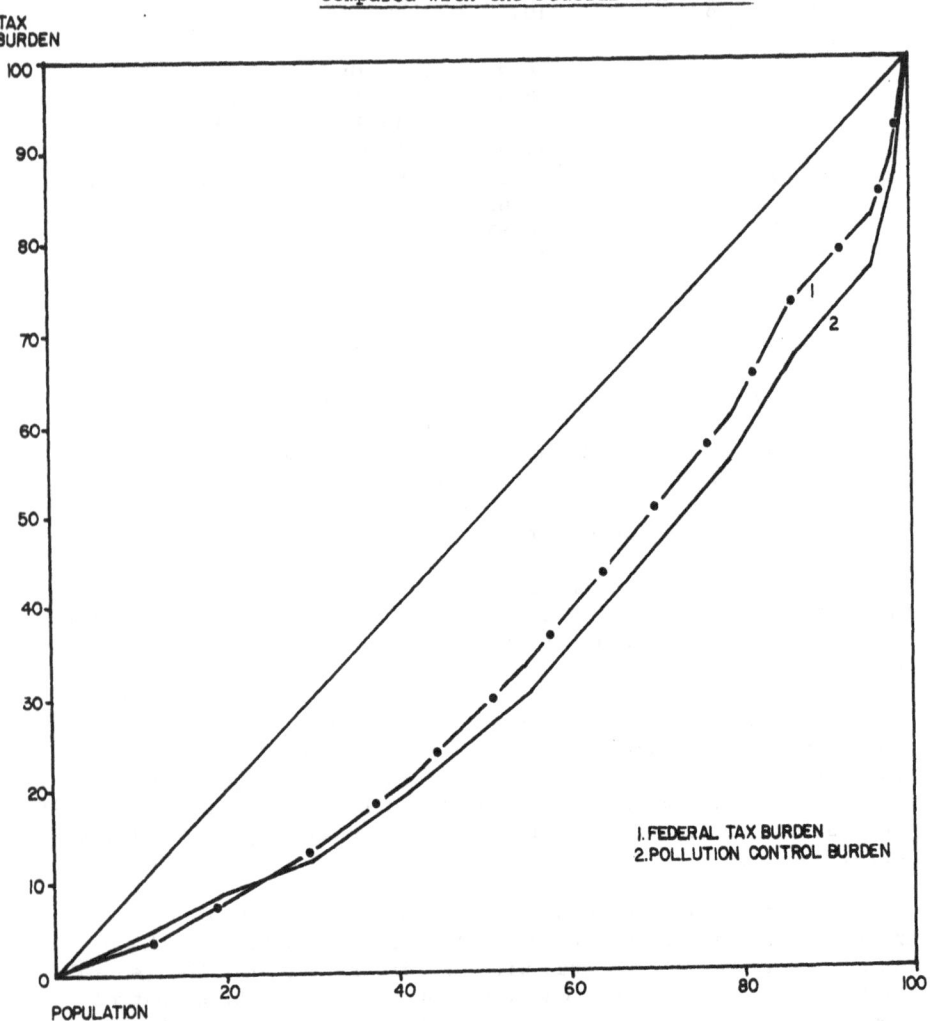

Categories I, II, and IVB, Zero Baselines, 1977

regressive over time. In absolute terms, the total burden grows from $4,226 million in 1977 to $4,900 million in 1980 and $4,534 million by 1985. During this period of time, as construction continues, annualized costs grow. The percentage distributions indicate that these greater costs are distributed more regressively in 1980 than they were in 1977, and still more regressively in 1985. As more facilities are constructed during the period, the associated O&M charges increase as well. Further, O&M charges represent an increasing proportion of the total annual costs, because, while O&M costs grow in equal annual increments, annual capital cost increments are calculated through the sinking fund formula. Since O&M charges are financed through the most regressive revenue instruments--user charges and property taxes--, it might be expected that the distribution of the burden will shift towards the lower income groups. According to our assumption, the amount of construction assumed by the 1974 "Needs Survey" is completed by 1983; therefore, the amount of O&M expenditures does not increase after that time. This in turn implies that the distribution of the burden remains constant after 1983. Exhibit III-74 presents Lorenz Curves for 1977, 1980, and 1985, in order to compare the distribution of burdens over time.

Regional differentials may also be seen in these tables. The costs, in absolute terms, are the highest in the South and progressively lower in the Northeast, the North Central and the West. This result is simply based on the distribution of the "needs" obtained in the Needs Survey. The percentage distribution of the burdens is the most progressive in the Northeast and the most regressive in the South. In 1985, for example, the lowest two income groups in the South are responsible for 11.7% of the burden; the comparable percentage for the Northeast is 5.0%. The reason for these differentials may be found in the underlying distribution of income; in the South a much larger (26.8%) fraction of the population falls into the lowest two income groups than in the Northeast (14.7%). Another hypothesis is that in the South, more regressive tax instruments are used than in the Northeast. Exhibit III-75 presents Lorenz curves for the South and the Northeast for 1985.

Needs Survey Categories I-V, with Zero Baselines, for 1985

Exhibit III-76 describes the distribution of the cost burdens for Needs Survey categories I-V. This table may be compared to Exhibit III-71 which presents the distribution of costs for 1985 for categories I, II, and IVB. The major difference between the two scenarios is the amount of expenditures: for 1985, the total annual cost for categories I, II, and IVB is $5,534 million, and for categories I-V it is $11,850 million. Per family costs for categories I-V are described in Exhibit III-77. The distribution of the costs by income group for categories I, II, and IVB are more regressive than the distribution of the costs of categories I-V. The reason for this difference may be explained in terms of

Exhibit III-74

Lorenz Curves: The Distribution of Pollution Control Burdens
for 1977, 1980, and 1985: Categories I, II, and IVB,
Zero Baselines

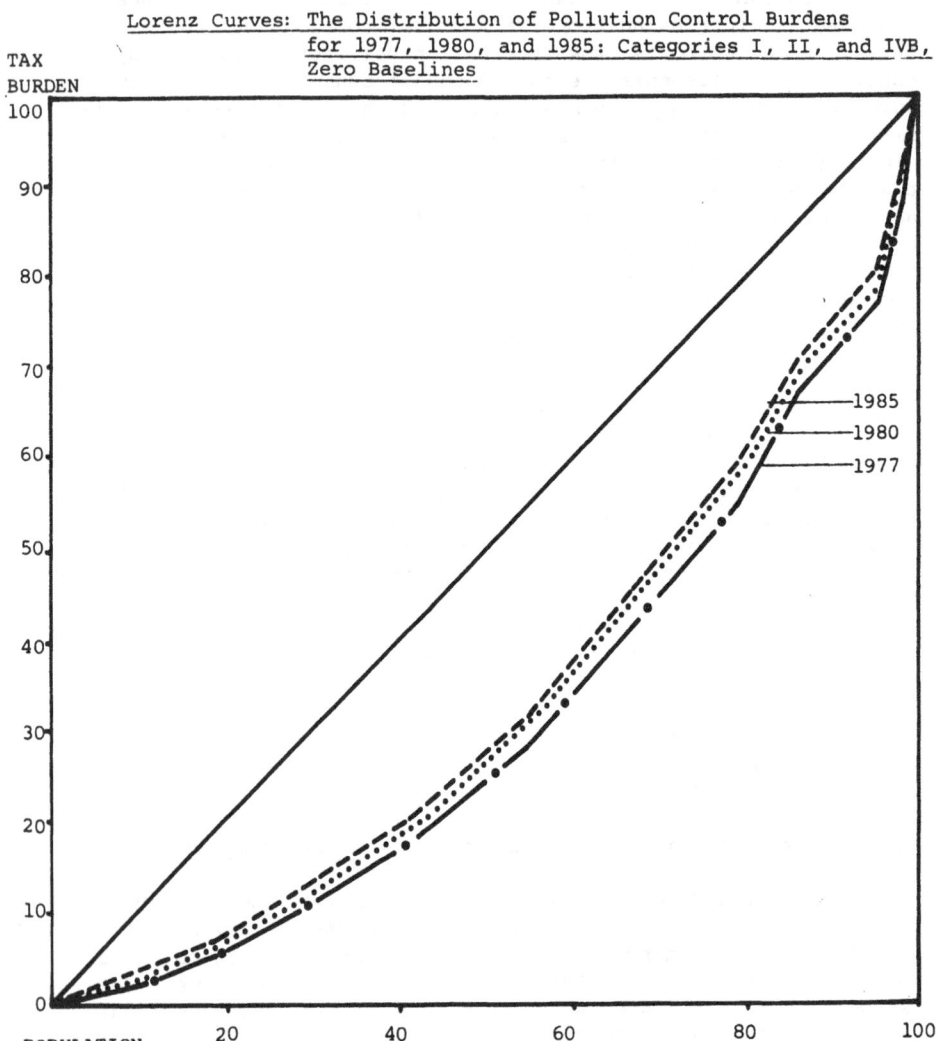

EXHIBIT III-75

Lorenz Curves: The Distribution of Pollution Control
Burdens in the Northeast and the South:
Categories I, II, and IVB, Zero Baselines,
1985

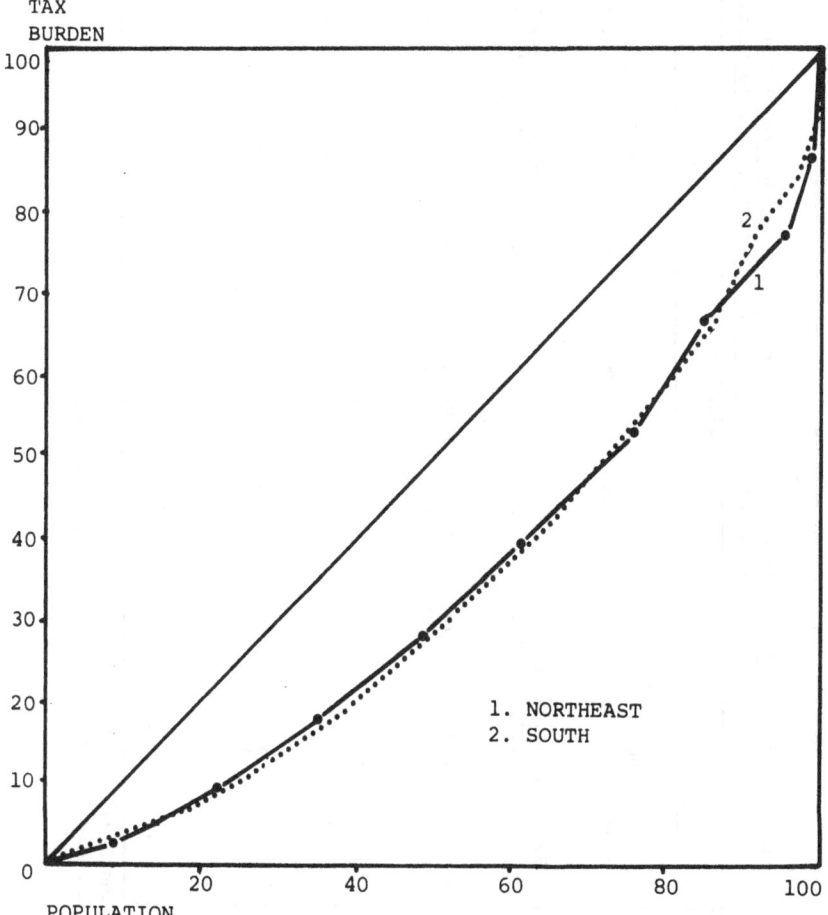

1. NORTHEAST
2. SOUTH

Exhibit III-76

1985 Distribution of the Costs of Publicly-Owned Wastewater Treatment Facilities by Income Group, for the U.S. and Census Regions: Categories I through V -- Zero Baselines

REGION	INCOME GROUP									
	1	2	3	4	5	6	7	8	9	10
UNITED STATES										
Total ($Millions)*	$391.8	$367.8	$645.3	$1024.0	$1337.1	$3028.8	$1575.7	$1107.3	$1024.6	$1347.3
Percent of Total**	3.3%	3.1%	5.4%	8.6%	11.3%	25.6%	13.3%	9.3%	8.6%	11.4%
NORTHEAST										
Total ($Millions)	$ 77.4	$ 81.6	$158.8	$ 271.2	$ 373.8	$ 900.7	$ 508.5	$ 361.0	$ 326.1	$ 501.2
Percent of Totals	2.2%	2.3%	4.4%	7.5%	10.4%	25.0%	14.1%	10.0%	10.1%	13.9%
NORTH CENTRAL										
Total ($Millions)	$108.3	$103.2	$185.5	$ 317.7	$ 451.1	$1065.0	$ 545.2	$ 382.6	$ 318.0	$ 401.2
Percent of Totals	2.3%	2.7%	4.8%	8.2%	11.6%	27.5%	14.1%	9.9%	8.2%	10.3%
SOUTH										
Total ($Millions)	$164.5	$140.8	$226.6	$ 316.0	$ 356.5	$ 688.2	$ 309.1	$ 214.6	$ 199.9	$ 272.7
Percent of Total	5.7%	4.9%	7.8%	10.9%	12.3%	23.8%	10.7%	7.4%	6.9%	9.4%
WEST										
Total ($Millions)	$ 36.2	$ 37.0	$ 63.7	$ 103.8	$ 138.1	$ 329.7	$ 178.7	$ 124.0	$ 114.3	$ 139.7
Percent of Totals	2.9%	2.9%	5.0%	8.2%	10.9%	26.1%	14.1%	9.8%	9.0%	11.0%
ALA. HAW., D.C.										
Total ($Millions)	$ 5.4	$ 5.2	$ 10.6	$ 15.5	$ 17.6	$ 45.4	$ 34.1	$ 25.1	$ 30.3	$ 32.5
Percent of Totals	2.4%	2.4%	4.8%	7.0%	7.9%	20.5%	15.4%	11.3%	13.7%	14.7%

*Costs in Millions $1973

**Percent of each region's total expenditures borne by each income group.

174

Exhibit III-77

Annual Per Family Costs for Needs Survey Categories I-V

in 1985

Income Group	Per Family Cost $1973	Per Family Cost as % of Income
1	$ 45.29	1.15%
2	74.60	0.78%
3	93.66	0.71%
4	121.47	0.68%
5	132.91	0.60%
6	175.79	0.60%
7	287.54	0.74%
8	185.17	0.33%
9	485.49	0.15%
10	2405.89	0.52%

our assumptions concerning O&M costs. For categories I-V, the ratio of O&M to capital costs was assumed to be approximately half as large as the ratio for categories I, II, and IVB. This assumption reflects the fact that categories I-V include collection systems, which typically have low O&M costs.

Needs Survey Categories I, II, and IVB, with High Baselines, 1985

Even without the passage of the Water Pollution Control Act of 1972, some (baseline) water pollution control expenditures would have been incurred. P.L. 92-500, however, imposed additional expenditures; in this scenario the distribution of the incremental expenditures are examined.

The first step in these calculations was estimating the distribution of the costs for categories I, II, and IVB, as has been described above. Second, the distribution of the baseline expenditure was calculated. Third, for each income group, the burden of the baseline expenditures was netted from the total cost burden of categories I, II, and IVB. The resulting difference describes the distribution of the incremental costs imposed by the Act. These results are presented in Exhibit III-78.

According to this scenario, P.L. 92-500 imposes $3,136 million of additional costs for 1985; in its absence, expenditures would have been $2,399 million. The distribution of the high baseline expenditures is described below in Exhibit III-79.

Exhibit III-78

1985 Distribution of the Incremental Costs of Publicly-Owned Wastewater Treatment Facilities by Income Group, for the U.S. and Census Regions: Categories I, II, and IVB -- High Baselines

REGION	INCOME GROUP									
	1	2	3	4	5	6	7	8	9	10
UNITED STATES										
Total ($Millions)*	$127.9	$107.4	$188.5	$282.3	$351.2	$772.8	$386.0	$286.5	$274.6	$358.8
Percent of Total**	4.1%	3.4%	6.0%	9.0%	11.2%	24.6%	12.3%	9.1%	8.8%	11.4%
NORTHEAST										
Total ($Millions)	$ 5.5	$ 6.0	$ 20.5	$ 32.2	$ 56.0	$150.7	$ 87.1	$ 68.7	$ 77.3	$110.7
Percent of Total	0.9%	1.0%	3.3%	6.1%	9.6%	24.3%	14.0%	11.1%	12.5%	17.8%
NORTH CENTRAL										
Total ($Millions)	$ 16.1	$ 15.7	$ 31.2	$ 52.8	$ 74.4	$178.3	$ 87.9	$ 65.6	$ 61.2	$ 81.7
Percent of Total	2.4%	2.3%	4.7%	8.0%	11.2%	26.8%	13.2%	9.9%	9.2%	12.3%
SOUTH										
Total ($Millions)	$ 88.9	$ 70.0	$108.4	$147.0	$163.4	$304.6	$130.9	$ 93.2	$ 81.5	$106.4
Percent of Total	6.9%	5.4%	8.9%	11.4%	12.6%	23.5%	10.1%	7.2%	6.3%	8.2%
WEST										
Total ($Millions)	$ 15.6	$ 14.6	$ 24.9	$ 39.3	$ 51.4	$122.3	$ 66.2	$ 48.6	$ 43.7	$ 51.5
Percent of Total	3.3%	3.1%	5.2%	8.2%	10.8%	25.6%	13.9%	10.2%	9.1%	10.8%
ALA., HAW., D.C.										
Total ($Millions)	$ 1.8	$ 1.7	$ 3.5	$ 5.0	$ 6.0	$ 16.9	$ 13.9	$ 10.4	$ 10.9	$ 8.5
Percent of Total	2.3%	2.2%	4.5%	6.4%	7.6%	21.5%	17.7%	13.2%	13.9%	10.8%

*Costs in Millions of $1973.
**Percent of each region's total expenditures borne by each income group.

Exhibit III-79

The Distribution of High Baseline Expenditures

Income Group	Baseline Cost Burden in $Millions	Percent of Baseline Cost Burden Borne by Each Income Group
1	$103.8	4.3%
2	91.6	3.8%
3	145.9	6.1%
4	222.9	9.3%
5	288.8	12.0%
6	624.9	26.0%
7	314.9	13.1%
8	220.5	9.2%
9	176.8	7.4%
10	209.0	8.7%

Exhibit III-79 shows that the distribution of the high baseline expenditures is far more regressive than the distribution of burdens in any other scenario. Without the Act, federal participation in the construction grant program would have been more limited than it is at the present time. As local tax instruments are more regressive than federal taxes, baseline expenditures may be expected to be regressive.

Per family costs with and without the Act, as well as the incremental expenses caused by the Act, are described in Exhibit III-80.

The distribution of the incremental expenditures imposed by the Act, as presented in Exhibit III-78, is a more progressive distribution than the distribution in other scenarios; the low income groups, paying a substantial proportion of costs in the baseline case, pay a smaller proportion of the additional costs, while the opposite is true for the high income categories.

Exhibit III-81 presents Lorenz curves for the distribution of baseline expenditures, of the incremental expenditures, and of the total expenditures. As may be expected, the baseline expenditures are the most regressive, and the incremental expenditures are the most progressive. Indeed, the incidence of the total expenditures may be thought of as a weighted average of these. In other words, the impact of the Act is not limited to increasing the expenditures necessary for achieving its requirements, but to shift the relative distribution of the burden from the lower to the higher income groups.

Exhibit III-80

Annual Per Family Costs With and Without the Act, and
Incremental Costs, 1985

Income Group	($1973)			(As a % of Income)		
	Baseline Costs	Incremental Costs	Total Costs	Baseline Costs	Incremental Costs	Total Costs
1	$12.00	$14.79	$26.79	0.31%	0.37%	$0.68%
2	18.58	21.78	40.36	0.19%	0.23%	0.42%
3	21.18	27.36	48.54	0.16%	0.21%	0.37%
4	26.44	33.49	59.93	0.15%	0.19%	0.34%
5	28.70	34.91	63.61	0.13%	0.16%	0.29%
6	36.27	44.85	81.12	0.13%	0.15%	0.28%
7	57.46	70.44	127.94	0.15%	0.18%	0.33%
8	36.87	47.91	84.78	0.06%	0.09%	0.15%
9	83.79	130.14	213.93	0.07%	0.10%	0.17%
10	373.21	640.17	1013.38	0.08%	0.14%	0.22%

Exhibit III-81

Lorenz Curves: The Distribution of Burdens for Baseline
Expenditures, Incremental Expenditures
Imposed by P.L. 92-500, and Total Expenditures under P.L. 92-500: Categories I, II,
and IVB, 1985.

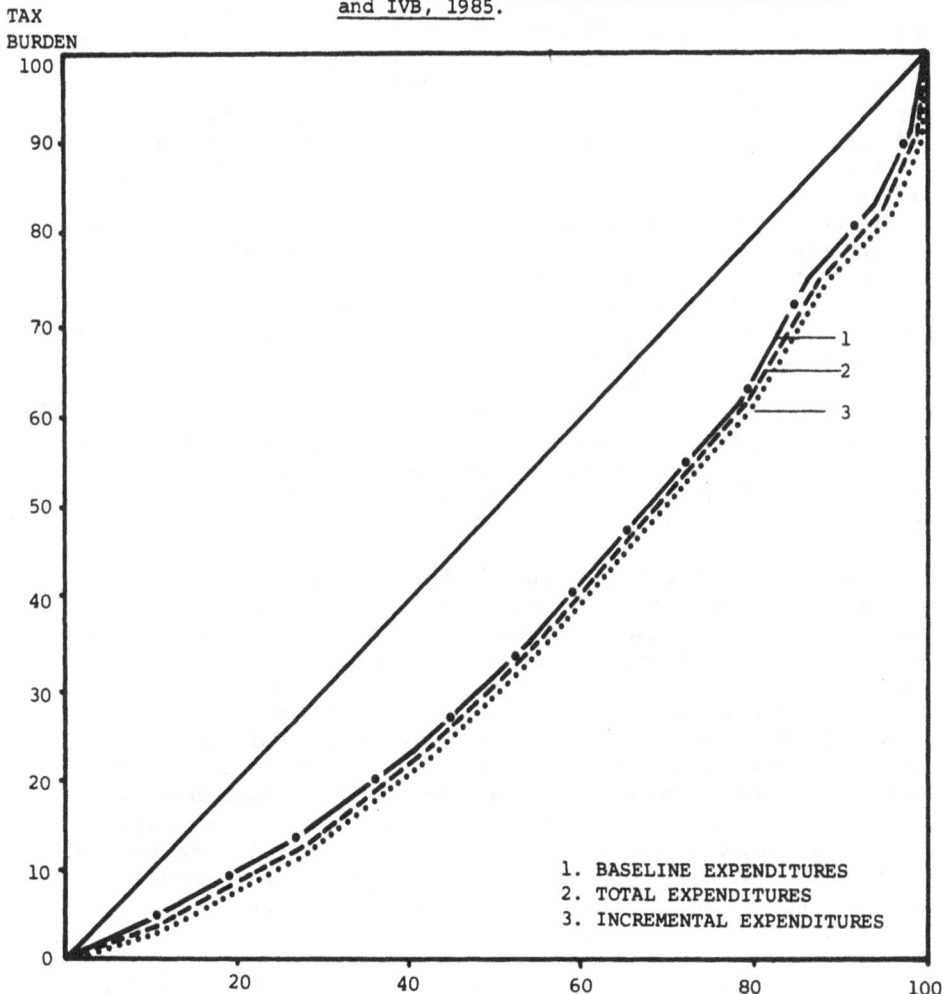

1. BASELINE EXPENDITURES
2. TOTAL EXPENDITURES
3. INCREMENTAL EXPENDITURES

179

Differences Between Counties, Special Districts, and Municipalities
in the Distribution of Cost Burdens

Pollution control expenditures are made by three types of
local governments: municipalities, counties, and special districts.
Of the $5,591 million to be spent in 1985 on categories I, II, and
IVB (assuming a zero baseline), some $3,780 million or 68% will be
spent by municipalities, $1,097 million (or 20%) will be spent by
special districts, and $674 million (or 12%) by county governments.
As different types of government rely on different fiscal instruments for financing their expenditures, the incidence of the burdens
may be expected to vary with the type of government responsible for
undertaking the necessary investments.   For example, special districts finance a relatively low proportion of their capital expenditures from the property tax; the national average is 27.3%.
According to our assumptions, however, the balance comes from intergovernmental transfers from other municipal governments.   As a significant fraction of these expenditures are financed through the
property tax, it may be anticipated that the incidence of the expenditures made by special districts would be more regressive than
the national average.   This is not the case, however, because expenditure reductions are ruled out for special districts according
to our assumptions.   Expenditure reductions, particularly on selected services (including parks, recreation, etc.) are extremely regressive.   The limitation of revenue sources to tax instruments
results in the more progressive distribution of burdens for special
districts.

Counties, on the other hand use the property tax fairly extensively, as was shown in Exhibit III-64.   The composition of their
tax revenues is reasonably similar to that of municipalities.   The
breakdown of expenditure reductions between highway expenditures and
expenditures on special services differs for counties and municipalities.*   Counties reduce their expenditures on these two categories
in roughly the same proportions, while most municipal expenditure
cuts fall into the selected services category.   As these expenditure
cuts are by far the most regressive, the municipal incidence pattern
is more regressive than the county incidence pattern.   Municipalities,
for example, rely on the property tax more extensively than do counties and special districts; therefore, the incidence of municipal
expenditures may be expected to be the most regressive.

The distribution of the costs by type of government is presented in Exhibit III-82.   According to this table, the distribution
of the burdens for counties and special districts are quite similar,
and more progressive than the incidence of municipal expenditures.

---

*We have assumed that the distribution of the benefits of ex-
penditures on selected services is the same as the distribution of
total expenditure benefits.   See Exhibit III-32.

180

Exhibit III-82

The Distribution of Costs of Publicly Owned
Wastewater Treatment Facilities by Income Group,
For Types of Governments: Categories I, II, and
IVB -- Zero Baselines

Type of Government	INCOME GROUP									
	1	2	3	4	5	6	7	8	9	10
MUNICIPALITIES										
Total ($Millions)*	$170.8	$145.2	$241.3	$358.8	$446.5	$952.4	$461.0	$332.9	$292.7	$378.2
Percent of Total**	4.5%	3.8%	6.4%	9.5%	11.8%	25.2%	12.2%	8.8%	7.7%	10.0%
SPECIAL DISTRICTS										
Total ($Millions)	$ 37.5	$ 33.5	$ 57.9	$ 92.6	$123.5	$283.5	$149.1	$108.1	$ 96.9	$114.7
Percent of Total	3.7%	3.3%	5.8%	9.3%	12.3%	28.3%	14.9%	10.8%	9.6%	11.5%
COUNTIES										
Total ($Millions)	$ 24.2	$ 21.0	$ 36.3	$ 55.6	$ 72.2	$166.1	$ 92.7	$ 67.1	$ 62.6	$ 76.0
Percent of Total	3.6%	3.1%	5.4%	8.3%	10.7%	24.7%	13.8%	10.0%	9.3%	11.3%

*Costs in Millions $1973.
**Percent of each governmental unit's total expenditures borned by each income group.

181

Exhibit III-83 presents Lorenz Curves for the distribution of burdens for municipalities and counties.

Differences in the Cost Burdens in Cities of Different Sizes

Of the $3,780 million spent by municipalities in 1985 on categories I, II, and IVB (assuming a zero baseline), cities with populations over 100,000 will spend $1,608 million, or 42%. Municipalities with populations under 5,000 will spend $275.5 million, or roughly 7%. The distribution of expenditures by city size and by income group for each city size is described in Exhibit III-84; the corresponding percentage distribution may be found in Exhibit III-85.

Different size cities may be expected to finance their expenditures from different revenue sources; in particular, the larger cities may be expected to draw on a wider tax base, and, therefore, have a more progressive distribution of burdens than the smaller centers. Further, the distribution of income varies by city size, reinforcing the same effect.

According to Exhibit III-85, this is generally the case. The distribution of burdens are compared for the largest and smallest city sizes in Exhibit III-86. The contribution of the property tax to the total tax burden does not, however, vary inversely with city size in each state. In Florida, for example, cities over 100,000 collect 56.52% of their tax revenues from the property tax, while centers under 5,000 collect only 48% through this method. Mississippi and Oklahoma are other examples.

In addition to the somewhat random pattern of the property tax, the structure of expenditure reductions varies with size of place. In the larger centers, expenditures (and expenditure cuts) on selected services are more important than the non-capital highway expenditures. In the smaller centers, for example, cities between 10,000 and 25,000, expenditures on services become relatively less important. Expenditure reductions then are more regressive in the larger centers than in the smaller ones; this helps to explain the relatively progressive incidence pattern for cities with populations between 10,000 and 25,000. For the two smaller size categories, the distribution of the population by income group is skewed towards the lower brackets. This accounts for their more regressive incidence patterns in spite of the progressive effect of the structure of expenditure cuts.

182

Exhibit III-83

Lorenz Curves: The Distribution of the Burden of Pollution
Control Expenditures made by Municipalities and Counties:
Categories I, II, and IVB, Zero Baselines, 1985

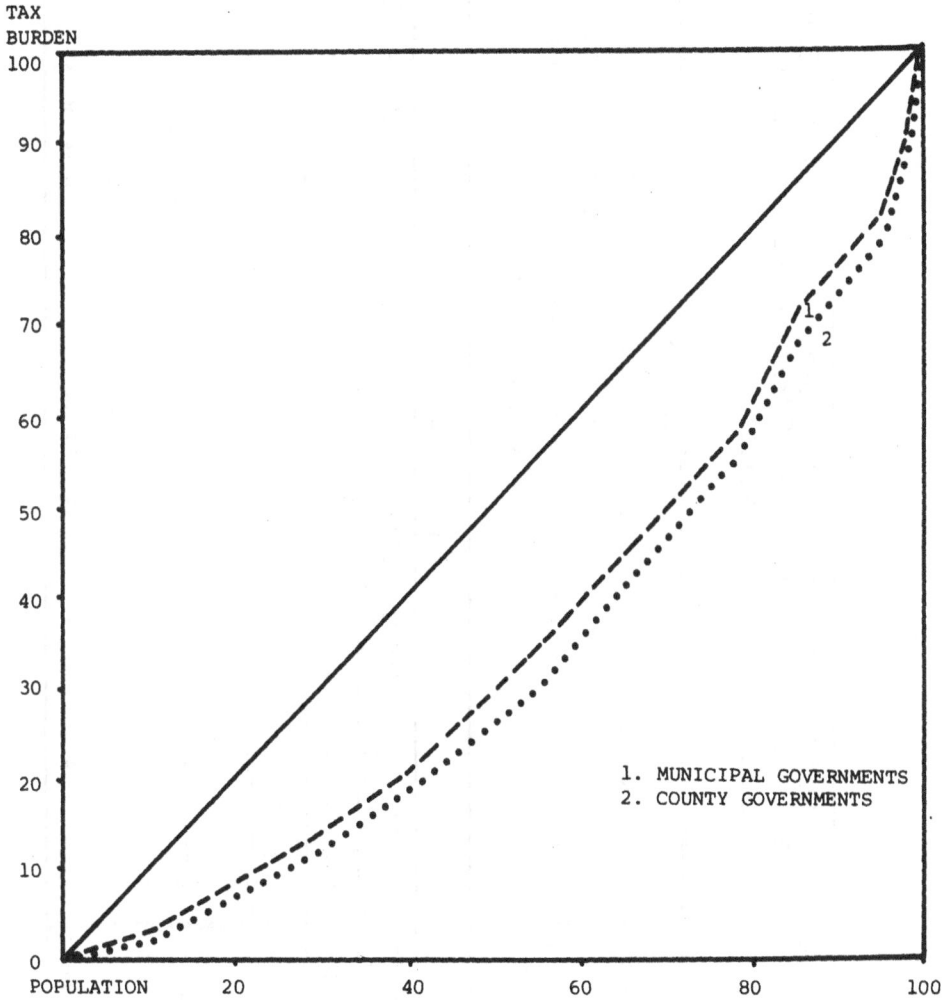

1. MUNICIPAL GOVERNMENTS
2. COUNTY GOVERNMENTS

Exhibit III-84

Impact of Different City Sizes on the Dollar Distribution of the 1985 Costs of Publicly-Owned Wastewater Treatment Facilities by Income Group: Categories I, II, and IVB, Zero Baselines

(Costs in millions of $1973)

City Size	1	2	3	4	5	6	7	8	9	10
100,000+	$70.4	$59.9	$100.1	$148.0	$184.2	$400.5	$201.2	$146.0	$130.2	$168.0
25,000-99,999	$45.7	$38.5	$63.3	$94.2	$116.9	$245.4	$115.0	$82.8	$71.8	$93.5
10,000-24,999	$26.2	$22.7	$38.1	$57.5	$72.7	$155.9	$75.0	$54.0	$47.3	$60.8
5,000-9,999	$14.8	$12.7	$21.4	$32.2	$40.2	$84.8	$40.1	$28.7	$25.2	$32.2
less than 5,000	$14.5	$12.2	$20.0	$29.3	$35.4	$71.4	$32.2	$22.9	$19.7	$25.5

Exhibit III-85

Impact of Different City Sizes on the Percentage Distribution of the 1985 Costs of Publicly-Owned Wastewater Treatment Facilities by Income Group: Categories I, II, and IVB, Zero Baselines

(Percent of each city size's total expenditures borne by each income group)

City Size	1	2	3	4	5	6	7	8	9	10
100,000+	4.4%	3.7%	6.2%	9.2%	11.4%	24.9%	12.5%	9.1%	8.1%	10.4%
25,000-99,999	4.7%	4.0%	6.5%	9.7%	12.1%	25.4%	11.9%	8.6%	7.4%	9.7%
10,000-24,999	4.3%	3.7%	6.2%	9.4%	11.9%	25.5%	12.3%	8.8%	7.8%	10.0%
5,000-9,999	4.4%	3.8%	6.5%	9.7%	12.1%	25.5%	12.1%	8.6%	7.6%	9.7%
less than 5,000	5.1%	4.3%	7.1%	10.3%	12.5%	25.3%	11.4%	8.1%	7.0%	9.0%

Exhibit III-86

Lorenz Curves: The Distribution of Pollution Control
Expenditures Made by Small and Large Cities: Categories
I, II, and IVB, Zero Baselines, 1985

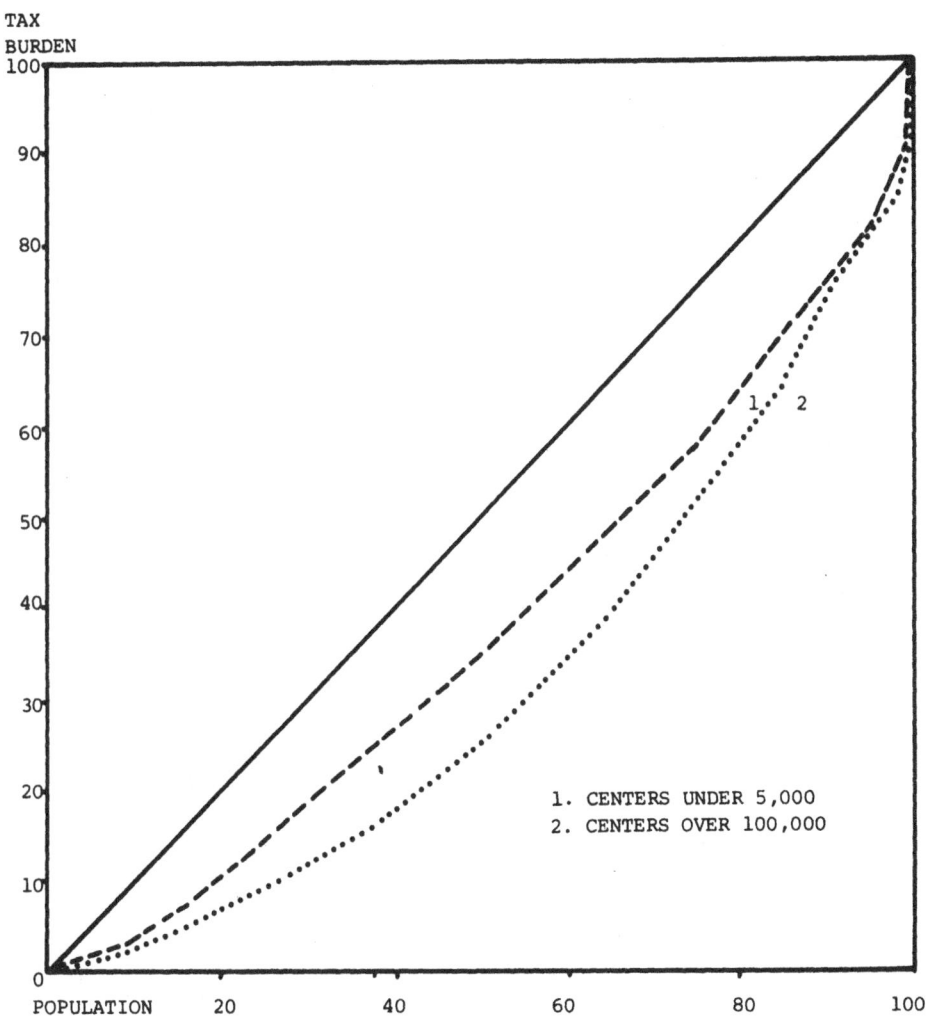

1. CENTERS UNDER 5,000
2. CENTERS OVER 100,000

185

IV. The Incidence of Industrial Price Increases

The increasingly stringent treatment requirements of
P.L. 92-500 apply to all point source dischargers, including
industries as well as municipalities. Industries may reduce
their wastes through self-treatment or through the use of
publicly owned treatment facilities. If they opt for the
latter alternative under current legislation, they must
generally meet pretreatment requirements, pay their
proportionate share under an industrial cost recovery scheme,
and contribute proportionally to O&M costs through user
charge systems. Thus, under either alternative--self-
treatment or joint municipal treatment--the costs of
production increase. The increased costs, in turn, are at
least partially passed along to consumers in the form of
higher prices. This increase in the price level means that
consumers can buy fewer goods and services with a fixed
income; in other words, their real income is reduced.

The chapter begins with a general overview of methods of
estimating the distribution of burdens caused by the price
increases. The succeeding sections discuss different elements
of this process: the calculation of price increases;
consumption patterns by income groups; and, finally, the
measures of welfare losses used in this study. The results
of these calculations are presented in the final section.

METHOD FOR ESTIMATING THE INCIDENCE OF
INDUSTRIAL TREATMENT COSTS

The distribution of the burdens caused by the industrial
treatment requirements of the Act depends on:

- the magnitude of the cost and price increases;

- their pattern: the relative magnitude of the price
  increases for different goods; and

- the consumption patterns of various subgroups
  of society.

Each of these will be described briefly before discussing
them more fully later in this chapter.

Cost and Price Increases

As industries comply with the treatment requirements
of the Act, their production costs increase. The choice
between self-treatment and the use of municipal facilities
will undoubtedly be dictated by considerations of cost
minimization; in any case, the costs of production will
reflect the use of resources for waste disposal purposes.

Treatment costs vary between industries as effluent
characteristics, treatment requirements, and technologies
differ. For example, the electroplating industry effluents
are particularly hard to treat; this in turn implies
disproportionate clean-up costs. Other things being equal,
higher than average price increases may be expected for the
electroplating industry.

As in the case of municipalities, industries would have
undertaken some water pollution control programs even if the
Federal Water Pollution Control Act of 1972 had not been
passed. Such baseline costs may be netted from the total
industrial expenditures necessary to meet the requirements
of the Act; this method estimates the incremental costs
imposed on industries by P.L. 92-500. Alternatively, baseline
expenditures may be ignored. This scenario estimates the
total costs of meeting the requirements of P.L. 92-500 for
industries.

Cost increases caused by water pollution control
requirements have been studied in a number of National
Commission on Water Quality contracts; these studies have
substantially understated baseline expenditures. Since the
baseline figures are in doubt, this study looked at the
distributional impacts of the total industrial cost of
meeting the requirements of P.L. 92-500 without making a
separate analysis for the incremental costs above the baseline.

The Pattern of Price Increases

The magnitude and pattern of cost increases is a partial
determinant of the price increases. Other influential factors
include the market structure the firm faces for its products
and factors, as well as interindustry relationships.

The control a firm has over its factor/product markets (monopsony, monopoly, and oligopoly) determines the extent to which it will attempt to, and succeed in, shifting these costs. In a monopsonistic situation, the industry will attempt to pass the increased costs back to the factors of production by lowering their rates of return, such as wages. The more usual situation involves monopolistic/oligopolistic control: here firms will raise the prices of their products in order to recover their increased costs. Under conditions of imperfect competition, the price increase may not be bounded by one; in other words, price increases may exceed pollution control costs. For any given level of treatment, pollution control costs may be expected to be higher at the early stages, due to high start-up costs, lack of experience, and so on. If firms pass these costs along to consumers in total, prices may "stick" at these high levels because of market rigidities, even though treatment costs have fallen.

In a competitive situation, in the long run, it may be expected that full cost increases will be passed along in the form of higher prices.

Although most of the water pollution control expenditures are concentrated in a small number of sectors, their effects are felt through the prices of most goods and services. The reason for this is that the output of the affected sectors serves as an input for the remaining industries; thus, the impact of the cost increases is passed to the full spectrum of final consumer goods and services. The industries with the largest pollution control costs include paper and allied products, petroleum refining, leather tanning, the production of iron and steel, and so on. As iron and steel products are used as inputs in other sectors, such as the construction industry and automobile manufacturing, construction costs and automobile prices may be expected to increase.

Cost increases for each industry were factored through the INFORUM Input-Output model of the University of Maryland. The I/O model specifies the structural interindustry relationships in the economy; it quantifies the amount of input each unit of product requires from every other I/O product category. The Input-Output model projects price increases for sectors such as petroleum refining and wholesale and retail trade. Since these categories do not correspond to the categories of goods/services on which consumers spend their money, the I/O and Personal Consumption Expenditure (PCE) categories had to be reconciled. This was done by using a bridge matrix, which specifies the

189

industrial composition of each PCE category.* The result of
these calculations was an estimate of the expected price
increase for each PCE category.

For a simple example of the bridge matrix, consider the
PCE category "men's footwear." Assume that the production of
footwear requires three inputs: leather, rubber and canvas.
If the relative magnitudes of these inputs are 50%, 25%, and
25%, and the price increases for leather, rubber and canvas
are 3%, 2%, and 1%, respectively, the resulting price increase
in footwear is 2.25%. The price increase calculated through
the bridge matrix for each PCE category is a simple weighted
average, with the weights reflecting the industrial
composition of each sector.

Consumption Patterns

Consumption patterns vary with income, age, regional
location, urban/rural residence, and other factors. Not
only do consumers spend a varying proportion of their income
on goods and services, but the bundle of goods they select
vary in a systematic fashion. For example, families in the
lowest income brackets consume in excess of their income,
while many of those in the higher brackets save a significant
proportion. Further, poor families allocate a greater
proportion of their expenditures to necessities than do
their rich counterparts. The price increases caused by
the industrial pollution control requirements of the Act
reduce families' real income in proportion to their propensity
to consume, and according to the composition bundle of
goods/services they consume.

If all price changes were equal, their impact could be
measured through the value of total consumption for each
income group. However, if price increases vary, their effect
is weighted by their importance in the budget. For example,
a relatively slight increase in food prices can reduce real
income substantially, because of the large proportion of
income spent on food, particularly in the low income
categories. The effect of such a price increase is clearly
regressive. The consumption patterns of the different
income and other socioeconomic groups are described in the

---

*The necessary calculations were performed by CONSAD
Research Corporation under a separate contract with the NCWQ.
These calculations are based on the Bridge Matrix developed
by the Department of Commerce. The long form of the bridge
matrix used was obtained directly from the Department of
Commerce.

third section of this chapter. The methodology for measuring the welfare losses caused by the price increases on the basis of consumption patterns is described in the fourth section of this chapter.

Exhibit IV-1 depicts the incidence of an increase in the price of food. The figure shows the distribution of expenditures on food between the various income groups. Although the lower income groups purchase a relatively small fraction of the total food consumed, their (food) consumption represents a relatively high percentage of their income. Thus, although the distribution of increased expenditures on food is distributed "progressively" among the income groups, the distribution of burdens relative to their income is regressive.

PRICE INCREASES

Real income depends, among other things, on the price level. If prices increase, other things being equal, consumers' purchasing power and real income decrease, with welfare losses being dependent on both the magnitude and pattern of price increases. In this section, some theoretical points concerning price increases will be discussed first. Second, the method of calculating price increases by the industry studies and CONSAD will be reviewed briefly. Finally, the resulting patterns of price increases will be described.

Price Increases in Theory

In general, one would expect the industrial cost increases produced by P.L. 92-500 to increase the prices of goods and reduce their quantities. The relative effect of the cost increase on raising prices versus reducing quantities depends upon a number of factors.

Suppose the requirements of P.L. 92-500 imposed additional costs on only a few firms within any particular industry. In this case, unless the affected firms possessed some monopolistic power, we would expect them to leave the industry: their costs would now be higher than average industry costs, and in the absence of structural anomalies no price increase would be possible; profits of these few firms would fall, and the firms would leave. If entry is free and firms numerous, the industry price level would not be affected.

The more usual result of P.L. 92-500 is an industry-wide price increase. The situation here is somewhat different.

191

Exhibit IV-1

Distribution of Industrial Price Increases

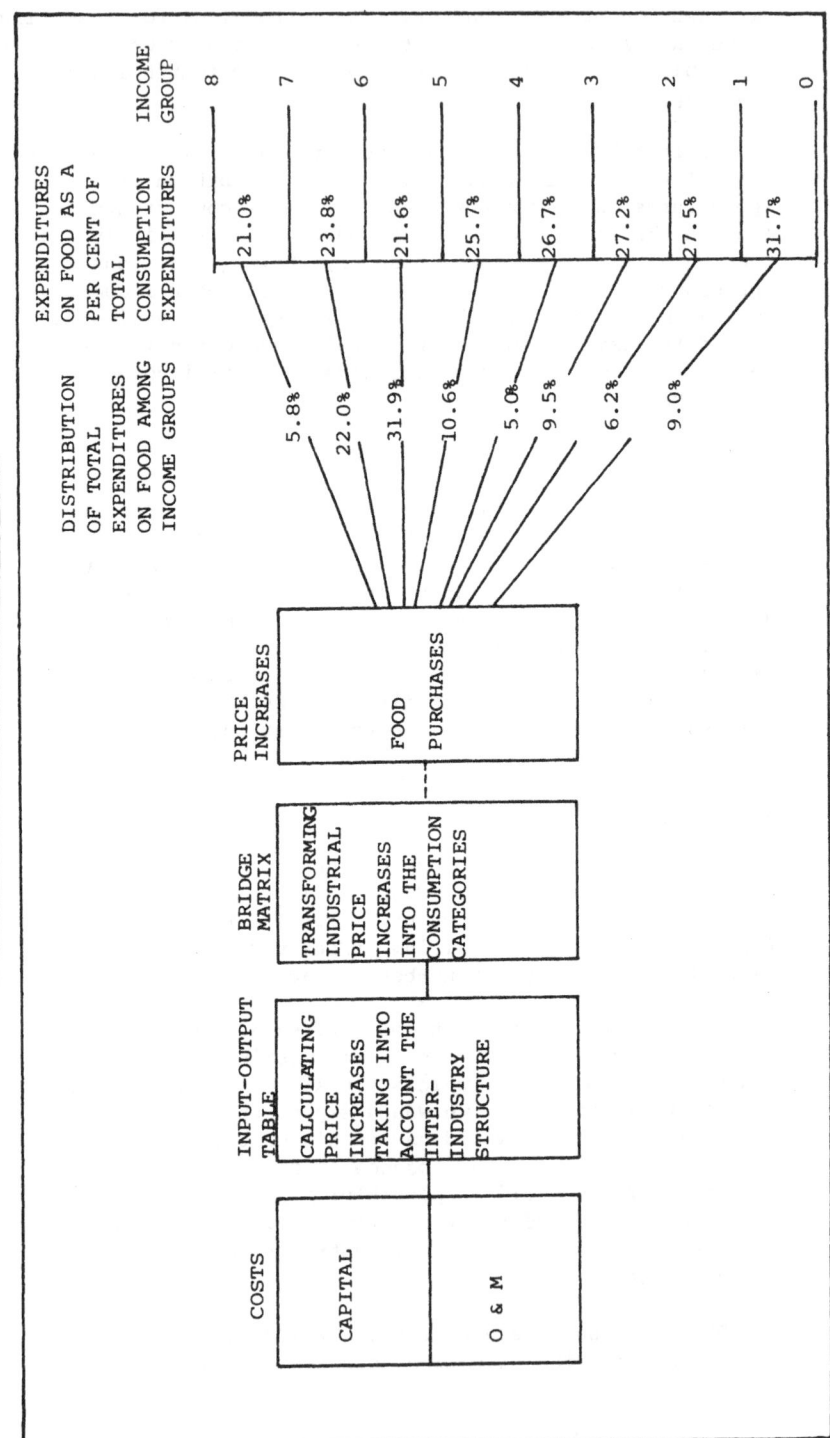

As a result of the cost increase, the industry supply curve
shifts to the left--i.e., it becomes costlier to supply any
given quantity of the good in question. The effect of this
shift on prices may be illustrated through the use of two '
polar examples. In Figure A below, we have assumed that
industry demand is perfectly inelastic: here prices rise by
the total amount of the cost increase. Conversely, if
industry demand is perfectly elastic (Figure B) prices remain
constant, but the quantity produced falls. The effect of the
cost increase on prices, then, depends upon elasticities of
demand.

Figure A   Figure B

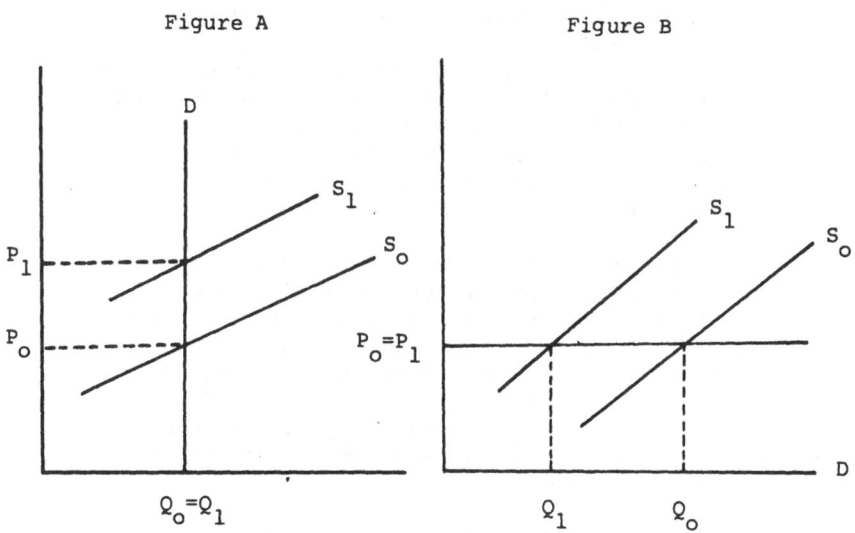

There is one special case in which the cost increases
may result in a price increase in excess of the cost increase.
Suppose cost increases are imposed on an oligopolistic
industry. It is often argued that such industries depend
upon external factors to "signal" the appropriate time for a
price increase. In this situation, oligopolies may use the
cost increase as a signalling device to coordinate pricing
policy and thus increase prices in excess of the cost
increase.*

Thus far, we have analyzed price increases in a partial
equilibrium framework. There are, however, some general
equilibrium considerations which come into play here.

———————————

*This analysis assumes that the oligopolistic firms were
not pricing "optimally" prior to the cost increase.

193

First, price increases in any one industry may result in cost increases in other industries. This effect is particularly strong if the affected industry produces an intermediate product--i.e., one that is used in the production of a second good. In this project, these general equilibrium effects were accounted for by the use of an Input-Output Model. It should be noted that this model assumes that no substitution occurs; that is, the receiving industry does not change its purchases as a result of the changes in the price of its factors of production.

Finally, the price increases produced by P.L. 92-500 may have some effect on the general level of consumption and savings. Income may be used for three alternative purposes: present consumption, future consumption, and accumulation. These are substitutes for one another, with the marginal rates of substitution dependent on individual preferences.* The individual preferences for these three alternative uses may be thought of in terms of a set of three dimensional preference surfaces.

In general, the price increases caused by the industrial treatment requirements of the Act may be considered analogous to a tax on consumption, discriminating against consumption in favor of accumulation, resulting in an increase in the rate of savings. This is indeed the expected outcome if the price increases are relevant (and equal) for the present and the future. The actual level of consumption, and possibly savings, will decrease, due to the negative "income effects" of the price rise.

If, on the other hand, the price increases are relevant for the future only, they are equivalent to a tax on future consumption. Such a tax discriminates against future consumption, in favor of both accumulation and present consumption. Thus, the impact on savings is both positive and negative--with the outcome dependent on the actual shape of the preference surface. A priori, we cannot predict whether the rate of savings will increase, decrease, or remain the same. The price rise will have the negative income effect, as discussed above. At the present time, most of the price increases related to water pollution control have not yet been realized; therefore, this second case is relevant.

---

*The discussion which follows is based on Sten Thore, Household Saving and the Price Level, National Institute of Economic Research, Stockholm, 1961.

Price Increase Calculations

The treatment requirements of the Act for industries have been described in Chapter I. These are BPT (best practicable treatment) for all point source dischargers by 1977, and BAT (best available technology) by 1983. BAT is also required for all new sources coming into production before 1983. Industries using publicly owned treatment facilities are subject to pretreatment requirements, which establish standards for wastes which are not compatible with municipal systems. To estimate the costs of meeting these requirements, and to estimate the resulting impacts on industries and on regional economies, the National Commission on Water Quality has undertaken a number of industry studies. Ten industries have been selected for in-depth study; these are:

- pulp and paper
- canned and preserved fruits and vegetables
- electroplating
- organic chemicals
- inorganic chemicals
- plastics and synthetics
- petroleum refining
- iron and steel
- textiles
- steam electric power.

The remaining industries were studied in somewhat less detail.

Industries have undertaken some limited pollution control programs in the past, and would have done so in the future even without the Act. Such "baseline" expenditures form part of their investment forecasts. The industry studies estimating the costs of meeting the BPT and BAT requirements of the Act did not explicitly take baseline expenditures into account. Some baseline abatement expenditures are included in the industrial investment forecasts which were made without regard for BPT and BAT requirements. In the forecast equations, the recent years are not given extra weights. Since most of the pollution control expenditures have been undertaken in recent years (in the early 1970's), the forecasts understate the abatement costs which would have been incurred without the Act. The cost estimates developed by the industry studies estimate the additional necessary

costs. These understate the total costs of pollution control
by the amount industries had planned to spend according to
their investment forecasts.

The technological alternatives studied were mainly
self-treatment or treatment in publicly owned facilities.
Process change as an alternative was largely ignored. To the
extent that process changes represent a potential for cost
savings, the cost estimates developed by the industry studies
are overestimated.

Relative price increases are then calculated by the price
model of the Strategic Environmental Assessment System (SEAS).
SEAS may be described as a collection of interdependent
forecasting models, which relate the national economy to the
generation of pollutant residuals and the associated costs of
pollution control. Its major components include a macroeconomic model developed by Chase Econometric Associates,
an input/output model (INFORUM) for 185 sectors developed
by the University of Maryland, and a commodity relative prices
forecast model, developed at the University of West Virginia.

The price model estimates the relative prices of
commodities; it interacts with Inforum to forecast long run
shifts in personal consumption which results from changes in
the relative price structure. The price model is a normal
cost model, using forecasts of long run average costs by
industry to predict the price of their products. Prices are
forecast then as the sum of unit material costs, unit capital
costs and unit labor costs.* The interindustry effects of
price increases for particular sectors are then analyzed
through the INFORUM** input/output model. The INFORUM model
was operated for the Commission by CONSAD Research
Corporation. The I/O model may best be described as a table
of coefficients, a_{ij}, where a_{ij} is defined as the units of
product i used in making one unit of product j. In the model,
the units used for measuring each product is one 1971 dollar's
worth. Interindustry relationships are defined for 185 I/O
sectors.[+] Price increases for sector i are translated into

---

*For a more detailed description of this model, see '
CONSAD Research Corp., Macroeconomic Impacts of P.L. 92-500,
Chapter 4.5.

**For a detailed description of this model, see Clopper
Almon, et al., 1985: Interindustry Forecasts of the American
Economy, Lexington, MA : Lexington Books, 1974.

+The I/O sectors in turn are composed of 4-digit Standard
Industrial Classification category industries, with each SIC
code relating to only one I/O code.

price increases for every other sector for which i serves as an input.

The result is a forecast of the price changes in index number form, with 1971 as 1.00 for each of the 185 sectors. The model generated such price changes for three scenarios: (1) the base case, which did not take BPT and BAT requirements into consideration; (2) SEAS 1, in which both BPT and BAT requirements are met according to schedule; and (3) BPT only.

The price forecasts, however, were for the 185 I/O sectors of the model. Since most studies of consumption patterns are in a form compatible with the 83 Personal Consumption Expenditure (PCE) categories, a transformation, or bridge, between the I/O sectors and PCE sectors was required.

Such a transformation has been developed by the Department of Commerce; the bridge matrix presents a cross classification of PCE by the 83 functional categories used in the national income and produce accounts, and the 4-digit SIC industrial categories. A simple example of calculating price increases for PCE categories has been described at the beginning of this chapter. The bridge matrix used was developed from data in the 1967 national accounts; it was supplied in the long form by the Department of Commerce. The calculations of price changes for PCE categories were performed by CONSAD, as part of their work for the Commission. The use of this bridge matrix implicitly assumes that the industrial composition of the PCE categories remains constant; that no substitutions occur. This is analogous to the assumption of constant proportions of the input/output model.

Price Increases Resulting from Pollution Control

The models described above calculated annual price changes for the PCE categories in the base case and for the SEAS scenario 1.* The price changes are in index number form with 1971 being 1.00 for all categories.

The base case forecasts refer to price increases relative to 1971 for the scenario without the impact of P.L. 92-500, while the SEAS 1 case forecasts indicate price increases again relative to 1971 for a scenario where BPT and BAT investments are made on time. For any given year, the difference between these two price indices describes the

---

*SEAS 1 assumes BPT and BAT expenditures will be made on schedule; the base case ignores such costs.

197

price increase caused by the water pollution control
requirements set forth in the Act. The difference between
the SEAS 1 price and the base case price index represents
the price increase which is caused by the pollution control
requirements of the Act. Exhibit IV-2 presents the resulting
price increases for 49* PCE categories. By 1977, the general
price level is expected to increase by approximately 1.4% as
a result of the Act. PCE Category #22, water and other
sanitary services, experience the largest price increase--
4.3%. Category #20, electricity gains the second largest
price increase--2.7%. By 1980, the increase in the general
price level is forecast at 2.0%, and by 1985 it is 2.9%.
Categories #22 and #20--water and other sanitary services,
and electricity--continue to experience the largest price
rises. By 1985, they increase by 8.5% and 6.0% respectively.

Price indices were also calculated for a scenario
(SEAS 4) which considers BPT costs alone. The difference
between the base case price indices and the SEAS 4 price
indices describes the increase in prices caused by BPT
expenditures alone. The resulting price increases for 1985
are presented in Exhibit IV-3. The increase in the general
price level due to BPT is expected to be approximately 1.8%
by 1985. This may be compared with the 2.9% increase caused
by BPT and BAT. Again, the greatest increases are expected
for categories #22 and #20 (water and sanitary services, and
electricity): 7.6% and 4.5% respectively.

In order to compare the structure of the price increases,
to find out whether price increases were experienced by the
same sectors, price increases in particular sectors were
related to the national average. The resulting ratios were
similar for the two price increase scenarios. Water and
other sanitary services was an exception, its price increase
was four times the national average under the BPT-alone
(SEAS 4) scenario, and three times the national average in the
BPT and BAT (SEAS 1) case. Electricity was 2.5 times and
2.0 times the national average in the SEAS 4 and SEAS 1
cases. The price increases for these two categories were
relatively larger for the SEAS 4 scenario than for the SEAS 1
case.

CONSUMPTION PATTERNS

In order to estimate the impacts of price increases for
specific groups, their consumption patterns must be analyzed.

---

*NOTE: 34 categories have been omitted from our analysis
for reasons described in the following section.

Exhibit IV-2

Price Increases Caused by Pollution Control (Both BPT and BAT)

Personal Consumption Expenditure Category	1977	1980	1985
1. Food purchased for off-premise consumption	0.0111	0.0202	0.0283
2. Purchased meals and beverages	0.0087	0.0159	0.0223
3. Tobacco products	0.0067	0.0126	0.0184
4. Shoes and other footwear	0.0082	0.0137	0.0237
5. Women's and children's clothing and accessories	0.0071	0.0141	0.0214
6. Men's and boy's clothing and accessories	0.0071	0.0140	0.0214
7. Cleaning, dyeing, pressing, storage & repair of garments	0.0087	0.0170	0.0261
8. Jewelry and watches	0.0123	0.0250	0.0369
9. Toilet articles and preparations	0.0107	0.0204	0.0324
10. Barbershops, beauty parlors and baths	0.0087	0.0170	0.0261
11. Owner-occupied non-farm dwellings	0.0104	0.0194	0.0268
12. Tenant-occupied non-farm dwellings	0.0112	0.0211	0.0305
13. Furniture including mattresses and bedsprings	0.0080	0.0155	0.0230
14. Kitchen and other household appliances	0.0145	0.0303	0.0443
15. China, glassware, tableware and utensils	0.0097	0.0196	0.0139
16. Other durable house furnishings	0.0124	0.0241	0.0364
17. Semi-durable house furnishings	0.0075	0.0148	0.0226
18. Cleaning and polishing preparations	0.0147	0.0263	0.0407
19. Stationery and writing supplies	0.0145	0.0234	0.0345
20. Electricity	0.0267	0.0555	0.0601
21. Gas	0.0071	0.0135	0.0202
22. Water and other sanitary services	0.0426	0.0726	0.0848
23. Telephone and telegraph	0.0066	0.0129	0.0177
24. Drug preparations and sundries	0.0130	0.0234	0.0340

Exhibit IV-2 (Cont'd)
Price Increases Caused by Pollution Control (Both BPT and BAT)

Personal Consumption Expenditure Category	1977	1980	1985
25. Ophthalmic products and orthopedic appliances	0.0264	0.0440	0.0581
26. Privately controlled hospitals and sanitariums	0.0061	0.0119	0.0176
27. New cars and net purchases of used cars	0.0100	0.0219	0.0313
28. Tires, tubes, accessories and parts	0.0111	0.0208	0.0315
29. Auto repair, greasing, washing, parking, storage & rental	0.0089	0.0178	0.0264
30. Gasoline and oil	0.0062	0.0141	0.0269
31. Street and electric railway and local bus	0.0062	0.0112	0.0160
32. Taxicabs	0.0062	0.0112	0.0160
33. Railway (commutation)	0.0113	0.0209	0.0265
34. Railway (excluding commutation)	0.0113	0.0209	0.0265
35. Intercity bus	0.0062	0.0112	0.0160
36. Airline	0.0063	0.0151	0.0243
37. Other	0.0072	0.0127	0.0168
38. Books and maps	0.0076	0.0134	0.0186
39. Magazines, newspapers and sheet music	0.0126	0.0201	0.0266
40. Nondurable toys and sport supplies	0.0099	0.0190	0.0271
41. Wheel goods, durable toys, sports equipment, boats and pleasure craft	0.0115	0.0237	0.0394
42. Radio and television receivers, records, and musical instruments	0.0105	0.0211	0.0309
43. Radio and television repair	0.0089	0.0170	0.0261
44. Flowers, seeds and potted plants	0.0096	0.0189	0.0281
45. Admissions to spectator amusements	0.0071	0.0140	0.0217
46. Clubs & fraternal organizations, except insurance	0.0065	0.0129	0.0185
47. Higher education	0.0065	0.0129	0.0185
48. Elementary and secondary schools	0.0065	0.0129	0.0185
49. Other	0.0065	0.0129	0.0185

In order to describe the impacts of price increases accurately, comprehensive and up-to-date information was necessary.

Several data sources were considered; including the 1960-61 Bureau of Labor Statistics (BLS) study,* and the annual consumer surveys of the Michigan Research Center.** The BLS study has detailed information on consumption patterns for all PCE categories by income group, by family size, age, geographical location and so on. This data describes consumption patterns for 1960-61. As consumption patterns have changed since that time, we attempted to find a more up-to-date description. BLS conducted a more recent survey in 1971-72; however, its results will not be available until 1976. The Michigan Survey Research Center's annual consumer surveys were considered as an alternative. The surveys provide up-to-date time series information on consumption patterns by income and other characteristics. However, only a few broad categories of goods are examined, such as housing, automobiles, and consumer durables. Because price increases due to pollution control may be expected for the entire spectrum of consumption categories, the information available in these surveys is not adequate for the purposes of this analysis.

A fairly recent and comprehensive consumer survey was conducted by Daniel Starch and Associates;[+] this was used as one of the main data sources for the consumption analysis of this study. The Starch survey examined consumption patterns by income and other characteristics for a large number of consumption categories, such as food purchased for home consumption, women's coats, women's suits, and so on. Food purchased for home consumption corresponds to PCE category #1, while women's coats and women's suits are subcategories of PCE category #5, women's and children's clothing and accessories. The correspondence between the Danial Starch categories and the official PCE categories was established with assistance from the Department of Commerce. The Starch study described consumption patterns in terms of frequencies; for example the number of women's coats bought by particular income groups during the survey period (one year). Transactions within categories (such as women's coats) were given equal weights. To estimate the dollar value of expenditures by income group

---

*U.S. Bureau of Labor Statistics, Consumer Expenditures and Income, BLS report #237093, Supplement B, Part A, 1966.

**Michigan Survey Research Center, Survey of Consumers 1971-72.

[+]D. Starch & Associates, Profile of U.S. Consumer Market Segments, Mamaroneck, NY, 1969.

```
                          Exhibit IV-3
              Price Changes with BPT Only, 1985
```

Personal Consumption Expenditure Category	Price Increases Caused by BPT**
1. Food purchased for off-premise consumption	.0184
2. Purchased meals and beverages	.0143
3. Tobacco products	.0119
4. Shoes and other footwear	.0143
5. Women's and children's clothing & accessories	.0129
6. Men's and boy's clothing & accessories	.0129
7. Cleaning, dyeing, pressing, storage and repair of garments	.0162
8. Jewlery and watches	.0235
9. Toilet articles and preparations	.0184
10. Barbershop, beauty parlors and baths	.0162
11. Owner-occupied non-farm dwellings	.0165
12. Tenant-occupies non-farm dwellings	.0191
13. Furniture including mattresses & bedsprings	.0146
14. Kitchen and otherhousehold appliances	.0284
15. China, glassware, tableware & utensils	.0174
16. Other durable house furnishings	.0225
17. Semi-durable house furnishings	.0134
18. Cleaning and polishing preparations	.0243
19. Stationery and writing supplies	.0227
20. Electricity	.0447
21. Gas	.0094
22. Water and other sanitary services	.0762
23. Telephone and telegraph	.0114
24. Drug preparations and sundries	.0216
25. Ophthalmic products and orthopedic appliances	.0283
26. Privately controlled hospitals & sanitariums	.0111

**Calculated by subtracting the base price index from the price index with BPT (Best Practical Technology).

Exhibit IV-3 (Cont'd)	
Price Changes in BPT Only, 1985	
Personal Consumption Expenditure Category	Price Increases Caused by BPT**
27. New cars and net purchases of used cars	.0197
28. Tires, tubes, accessories and parts	.0187
29. Auto repair, greasing, washing, parking, storage and rental	.0165
30. Gasoline and oil	.0107
31. Street & electric railway and local bus	.0087
32. Taxicabs	.0087
33. Railways (commutations)	.0192
34. Railways (excluding commutations)	.0192
35. Intercity bus	.0087
36. Airline	.0140
37. Other	.0127
38. Books and maps	.0119
39. Magazines, newspapers and sheet music	.0179
40. Nondurable toys and sport supplies	.0197
41. Wheel goods, durable toys, sports equipment, boats and pleasure craft	.0213
42. Radio & television, receivers, records, and musical instruments	.0197
43. Radio and television repair	.0162
44. Flowers, seeds and potted plants	.0185
45. Admissions to spectator amusements	.0136
46. Clubs and fraternal organizations, except insurance	.0119
47. Higher education	.0119
48. Elementary and secondary schools	.0119
49. Other	.0119

**Calculated by subtracting the base price index from the price index with BPT.

203

and PCE category, the aggregate expenditures on each PCE category, reported by the Department of Commerce*, were adjusted
in terms of the frequency distributions obtained from the
Daniel Starch study. For example, the Starch study indicated
that the lowest income group was responsible for 10% of the
expenditures on women's and children's clothing and accessories, and the fifth income group accounted for 10.5%. The
1968 total expenditures on this category amounted to $25,360
million, with the lowest income group spending $2,536 million
and the fifth another $2,663 million.

On the basis of the D. Starch survey, consumption patterns for 22 PCE groups were described. For another 24 categories, expenditures were calculated on the basis of income
elasticities. These in turn were estimated by Clopper Almon
from the consumption patterns of the 1960-61 BLS study.**
The PCE categories analyzed on the basis of the D. Starch
study and on the basis of the BLS study are listed and dif-
ferentiated in Exhibit IV-4.

Although income elasticity estimates were available in
the Almon study for PCE categories #20 and #21, electricity
and gas, they were not used because of the particular importance of these categories; income elasticities and consumption patterns for them were estimated on the basis of some
preliminary results of the 1971-72 BLS survey.[+] For category
#12, tenant occupied non-farm dwellings, the Starch survey
contained no information, nor did the Almon study estimate the
income elasticity. Therefore, we have estimated consumption
patterns for this category on the basis of the 1968-1969
Survey of Consumers.[++]

In sum, our approach was to piece together a description
of consumption patterns from the best sources available. The
resulting 49 PCE categories are presented in Exhibit IV-5.
Exhibit IV-5 describes the percentage distributions derived from these; in other words, it describes the com-

---

*Department of Commerce, Survey of Current Business,
July 1969.

**Clopper Almon, The American Economy to 1975, New York:
Harper and Row, 1966.

[+]U.S. Department of Labor, Bureau of Labor Statistics,
News, USDL 75-212, April 16,1975.

[++]Michigan Survey Research Center, op.cit., 1968-96. The
1968-69 survey was selected for the sake of consistency with
the other consumption data.

position of the budget for the eight income groups.

It should be noted that the Starch survey collected information for eight income groups, which do not correspond
with the ten income groups used in this study. Starch's groups are*:

(1) Under $3,000
(2) $3,000-4,999
(3) $5,000-6,999
(4) $7,000-7,999
(5) $8,000-9,999
(6) $10,000-14,999
(7) $15,000-24,999
(8) $25,000 and over

Consumption patterns and welfare losses due to price increases
were calculated for these income groups, and then allocated to
the ten income groups used in this research in order to
obtain consistency with the distribution of burdens of publicly-owned treatment plants. The relationship between the income groups used in this study and the eight Starch income
groups is described in Exhibit IV-6. The distribution of
families by income groups is assumed to be the same in the
Starch study and our analysis, as both are based on census.
The within group distribution is assumed to be constant; for
example, in the $3,000-5,000 Starch group, 10% of the families
are assumed to receive incomes between $3,000 and $3,200.
This assumption is no doubt an oversimplification. However,
in the absence of more detailed distributional information,
it was necessary for the sake of comparability.

Before turning to a description of the consumption patterns, it should be noted that such information was collected
for only 49 categories. Thirty-four categories were omitted.
Because the 49 categories analyzed made up approximately 80%
of the total personal consumption expenditures for 1968,** and
because relatively modest price increases are anticipated for
the categories omitted, it was felt that an analysis on the
basis of 49 categories would yield satisfactory measures of
incidence.

According to the INFORUM base case forecasts for personal consumption expenditures, the 49 categories selected for
analysis comprise approximately 81% of the total PCE forecast
for each year. The price increase forecasts indicate an

_____

*The income figures pertain to 1968 census money income.

**Department of Commerce, <u>Survey of Current Business</u>, 1969.

Exhibit IV-4

List of Consumer Goods/Services Analyzed
Department of Commerce Personal Consumption Expenditure (PCE) Categories

	PCE Category	Name		PCE Category	Name
1	I.1*	Food purchased for off-premise consumption (n.d.c.)	12	IV.2	Tenant-occupied nonfarm dwellings (including lodging houses)--space rental (s.)
2	I.2*	Purchased meals and beverages (n.d.c.)	13	V.1*	Furniture, including mattresses and bedsprings (d.c.)
3	I.5	Tobacco products (n.d.c.)	14	V.2*	Kitchen and other household appliances (d.c.)
4	II.1*	Shoes and other footwear (n.d.c.)	15	V.3*	China, glassware, tableware, and utensils (d.c.)
5	II.3.a*	Women's and children's clothing & accessories except footwear (n.d.c.)	16	V.4*	Other durable house furnishings (d.c.)
6	III.3.b*	Men's and boys clothing & accessories except footwear (n.d.c.)	17	V.5*	Semidurable house furnishings (n.d.c.)
7	II.5	Cleaning, dyeing, pressing, alteration, storage, and repair of garments including furs (in shops) not elsewhere classified (s.)	18	V.6	Cleaning and polishing preparations, and miscellaneous household supplies and paper products (n.d.c.)
8	II.7*	Jewelry and watches (d.c.)	19	V.7	Stationery and writing supplies (n.d.c.)
9	III.1*	Toilet articles and preparations (n.d.c.)	20	V.8.a.	Electricity (s.)
10	III.2	Barbershops, beauty parlors and baths (s.)	21	V.8.b	Gas (s.)
11	IV.1	Owner-occupied nonfarm dwellings --space rental value (s.)	22	V.8.c	Household utilities--water & other sanitary services (s.)

(d.c.) durable commodities   (n.d.c.) nondurable commodities   (s.) services

The categories marked "" have been analyzed on the basis of the D. Starch survey, and the balance on the basis of the BLS study.

Exhibit IV-4 (Cont'd)

List of Consumer Goods/Services Analyzed
Department of Commerce Personal Consumption Expenditures (PCE) Categories

PCE Category	Name	PCE Category	Name
23 V.9*	Telephone and Telegraph (s.)	37 VIII.3d	Other (s.)
24 VI.1	Drug preparations and sundries (n.d.c.)	38 IX.1*	Books and maps (d.c.)
25 VI.2	Opthalmic products and orthopedic appliances (d.c.)	39 IX.2	Magazines, newspapers, and sheet music (n.d.c.)
26 VI.6	Privately controlled hospitals and sanitariums (s.)	40 IX.3	Nondurable toys and sports supplies (n.d.c.)
27 VIII.1a*	New cars and net purchases of used cars (d.c.)	41 IX.4*	Wheel goods, durable toys, sports equipment, boats, and pleasure aircraft (d.c.)
28 VIII.1b*	Tires, tubes, accessories, and parts (d.c.)	42 IX.5*	Radio and television receivers, records, and musical instruments (d.c.)
29 VIII.1c	Automobile repair, greasing, washing, parking, storage, and rental (s.)	43 IX.6	Radio and television repair (s.)
30 VIII.1d	Gasoline and oil (n.d.c.)	44 IX.7*	Flowers, seeds, and potted plants (n.d.c.)
31 VIII.2a	Street and electric railway and local bus (s.)	45 IX.8a*	Admissions to spectator amusements --motion picture theaters (s.)
32 VIII.2b	Taxicabs (s.)	46 IX.9*	Clubs and fraternal organizations except insurance (s.)
33 VIII.2c	Railway (commutation) (s.)	47 X.1	Higher education (s.)
34 VIII.3a	Railway (excluding commutation) (s.)	48 X.2	Elementary & secondary schools (s.)
35 VIII.3b	Intercity bus (s.)	49 X.3	Other (s.)
36 VIII.3c	Airline (s.)		

(d.c.) durable commodities   (n.d.c.) nondurable commodities   (s.) services

The categories marked "" have been analyzed on the basis of the D. Starch survey, and the balance on the basis of the BLS study.

Exhibit IV-5

Percentage Distribution of Personal Consumption Expenditures for Different Income Groups

Personal Consumption Expenditure Category	Income Group								U.S. Total
	1	2	3	4	5	6	7	8	
1. Food purchased for off-premise consumption	27.0%	22.1%	21.3%	20.1%	19.8%	19.3%	18.2%	16.3%	18.8%
2. Purchased Meals and Beverages	4.7%	5.4%	5.9%	5.6%	5.9%	6.0%	5.6%	4.7%	5.4%
3. Tobacco Products	3.3%	2.7%	2.7%	2.3%	2.0%	1.8%	2.5%	1.1%	2.1%
4. Shoes and other footwear	2.7%	2.2%	1.9%	1.7%	1.6%	1.4%	1.3%	1.0%	1.5%
5. Women's and children's clothing and accessories	8.0%	7.0%	6.4%	5.7%	5.7%	5.5%	5.3%	4.4%	5.5%
6. Men's and boy's clothing and accessories	3.2%	3.3%	3.3%	3.0%	3.2%	3.3%	3.0%	2.1%	3.0%
7. Cleaning, dyeing, pressing, storage and repair of garments	0.0%	0.2%	0.4%	0.4%	0.6%	0.9%	1.5%	2.8%	0.9%
8. Jewelry and watches	0.6%	0.8%	0.8%	0.9%	0.7%	0.9%	0.9%	0.9%	0.8%
9. Toilet articles and preparations	0.9%	1.2%	1.3%	1.2%	1.2%	1.2%	2.1%	1.1%	1.3%
10. Barbershops, beauty parlors and baths	0.3%	0.4%	0.6%	0.6%	0.7%	0.9%	1.2%	1.7%	0.8%
11. Owner-occupied non-farm dwellings	10.7%	9.8%	10.5%	11.9%	12.4%	12.7%	11.4%	8.3%	10.9%
12. Tenant-occupied non-farm dwellings	4.8%	8.0%	6.5%	8.0%	5.9%	2.9%	2.0%	1.9%	3.7%

Exhibit IV-5   (cont'd)

Percentage Distribution of Personal Consumption Expenditures for Different Income Groups

Personal Consumption Expenditures Category	Income Group								U.S. Total
	1	2	3	4	5	6	7	8	
13. Furniture including mattresses and bedsprings	0.8%	1.1%	1.5%	1.5%	1.7%	2.0%	1.8%	1.7%	1.6%
14. Kitchen and other household appliances	1.6%	1.6%	1.9%	0.7%	1.8%	1.9%	1.9%	1.6%	1.7%
15. China, glassware, tableware and utensils	0.6%	0.8%	0.7%	0.7%	0.6%	0.8%	0.8%	0.8%	0.7%
16. Other durable house furnishings	1.4%	1.4%	1.5%	1.7%	1.8%	1.9%	1.9%	1.6%	1.7%
17. Semi-durable house furnishings	1.1%	1.3%	1.3%	1.3%	1.3%	1.3%	1.2%	1.0%	1.2%
18. Cleaning and polishing preparations	2.1%	1.5%	1.3%	1.1%	1.1%	1.0%	0.9%	0.8%	3.8%
19. Stationery and writing supplies	0.1%	0.2%	0.3%	0.3%	0.3%	0.4%	0.6%	0.9%	0.4%
20. Electricity	1.7%	2.8%	2.4%	3.2%	2.6%	2.2%	0.7%	0.7%	1.7%
21. Gas	0.9%	1.5%	1.3%	1.8%	1.4%	1.2%	0.4%	0.4%	1.0%
22. Water and other sanitary services	0.3%	0.4%	0.4%	0.4%	0.4%	0.5%	0.5%	0.7%	0.4%
23. Telephone and telegraph	3.4%	2.3%	2.1%	2.0%	1.9%	1.7%	1.5%	1.2%	1.8%
24. Drug preparations and sundries	1.3%	1.3%	1.3%	1.2%	1.9%	1.3%	1.4%	1.6%	1.3%

Exhibit IV-5 (cont'd)

Percentage Distribution of Personal Consumption Expenditures for Different Income Groups

Personal Consumption Expenditure Category	1	2	3	Income Group 4	5	6	7	8	U.S. Total
25. Ophthalmic products and orthopedic appliances	0.1%	0.2%	0.3%	0.3%	0.3%	0.4%	0.5%	0.7%	0.3%
26. Privately controlled hospitals and sanitariums	2.6%	2.6%	2.7%	2.6%	2.6%	2.8%	3.1%	3.7%	2.7%
27. New cars and net purchases of used cars	5.4%	6.5%	7.0%	7.2%	7.3%	7.8%	8.0%	7.5%	7.0%
28. Tires, tubes, accessories and parts	0.5%	0.8%	1.0%	1.0%	1.1%	1.1%	1.0%	0.9%	0.9%
29. Auto repair, greasing, washing, parking, storage and rental	0.5%	0.9%	1.1%	1.2%	1.4%	1.7%	2.3%	3.3%	1.6%
30. Gasoline and oil	3.2%	3.6%	3.8%	3.8%	3.9%	4.2%	5.0%	6.2%	4.1%
31. Street and electric railway and local bus	0.2%	0.2%	0.3%	0.3%	0.3%	0.3%	0.4%	0.5%	0.3%
32. Taxicabs	0.1%	0.1%	0.1%	0.1%	0.1%	0.2%	0.2%	0.3%	0.2%
33. Railway (commutation)	0.0%	0.0%	0.0%	0.0%	0.0%	0.0%	0.0%	0.0%	0.0%
34. Railways (excluding commutation)	0.0%	0.0%	0.0%	0.0%	0.0%	0.0%	0.0%	0.2%	0.0%
35. Intercity Bus	0.0%	0.0%	0.0%	0.1%	0.1%	0.1%	0.1%	0.3%	0.0%
36. Airline	0.0%	0.0%	0.0%	0.1%	0.1%	0.4%	0.7%	1.6%	0.4%

Exhibit IV-5 (cont'd)

Percentage Distribution of Personal Consumption Expenditures for Different Income Groups

Personal Consumption Expenditure Category	1	2	3	Income Group 4	5	6	7	8	U.S. Total
37. Other	0.0%	0.0%	0.0%	0.0%	0.0%	0.0%	0.0%	0.0%	0.0%
38. Books and maps	0.5%	0.5%	0.6%	0.5%	0.5%	0.7%	0.6%	0.6%	0.6%
39. Magazines, newspapers and sheet music	0.2%	0.4%	0.5%	0.5%	0.6%	0.8%	1.0%	1.5%	0.7%
40. Nondurable toys and sports supplies	2.0%	1.5%	1.2%	1.1%	1.0%	0.9%	0.9%	0.8%	3.6%
41. Wheel goods, durable toys, sports equipment, boats and pleasure craft	0.6%	0.6%	0.8%	0.9%	0.9%	1.0%	1.0%	1.0%	0.9%
42. Radio and television receivers, records, and musical instruments	1.6%	1.7%	1.8%	1.7%	1.8%	1.9%	1.8%	1.6%	1.7%
43. Radio and television repair	0.2%	0.3%	0.3%	0.2%	0.3%	0.3%	0.3%	0.4%	0.3%
44. Flowers, seeds and potted plansts	0.3%	0.2%	0.2%	0.3%	0.3%	0.3%	0.3%	0.3%	0.2%
45. Admissions to spectator amusements	0.2%	0.2%	0.2%	0.2%	0.2%	0.2%	0.2%	0.2%	0.2%
46. Clubs and fraternal organizations except insurance	0.0%	0.0%	0.0%	0.1%	0.1%	0.2%	0.4%	0.7%	0.2%
47. Higher Education	0.0%	0.0%	0.0%	0.1%	0.2%	0.7%	1.6%	4.2%	0.9%
48. Higher Education	0.0%	0.0%	0.0%	0.0%	0.0%	0.4%	0.8%	2.1%	0.5%
49. Other	0.0%	0.0%	0.0%	0.0%	0.1%	0.4%	0.8%	2.2%	0.5%

Exhibit IV-6

Correspondence Between USR&E and Starch Income Categories	
USR&E Income Groups*	Starch Income Groups**
USR&E 1	S1 + 60% S2
USR&E 2	40% S2 + 40% S3
USR&E 3	60% S3 + 75% S4
USR&E 4	25% S4 + S5 + 5% S6
USR&E 5	41% S6
USR&E 6	55% S6 + 22% S7
USR&E 7	50% S7
USR&E 8	28% S7 + 12% S8
USR&E 9	68% S8
USR&E 10	20% S8

*USR&E income categories refer to the 10 income brackets used in this analysis.

**S1-S8 refer to the 8 income categories used in the D. Starch consumer survey.

average price increase of 2.7% by 1985 for the omitted categories and 3.2% for the categories analyzed. As the omitted categories consist largely of services, the forecast price increases are higher than anticipated. They may be explained, however, in terms of the indirect relationships which dominate the model. The exclusion of the thirty-four categories will understate the welfare losses resulting from price increases by less than 20%. As these categories are mainly services, their inclusion would represent additional burdens for the higher income groups. In other words, their exclusion introduces a regressive bias for our estimates.

The consumption patterns of our model largely conform to expectations. The average propensities to consume decline from 2.05 for the lowest income group to 0.05 for the highest. Propensities to consume are described in Exhibit IV-7. Note that the estimates of average propensities to consume have been calculated on the basis of expenditures for the 49 categories analyzed. Therefore, the propensity to consume is understated--the real propensity would be less for the lower income groups and more for the upper income categories. As expected, the "necessities" comprise larger proportions of the budget in the lower income groups than in the budgets of the higher income categories. The importance of food, tobacco, clothing and housing in the budget declines as income increases. Other expenditure categories such as automobile purchases, gas

212

Exhibit IV-7

Average Propensity to Consume by Income Group	
Income Group	Propensity to Consume
1	2.05
2	1.34
3	0.97
4	0.86
5	0.83
6	0.81
7	0.51
8	0.52

and oil, and airline transportation account for an increasing share of the budget for the higher income groups. Still other PCE categories, such as the durable and semi-durable house furnishings, are approximately of equal importance in the budgets of all income groups.

Changes in consumption patterns over time are described in Exhibits IV-8 and IV-9. Exhibit IV-8 shows the 1960 and 1970 budget proportions for selected PCE categories as calculated by the Conference Board. Similar budget proportions for 1968 and forecasts for 1985 are presented in Exhibit IV-9. The 1985 forecasts are based on income inelasticities. The changes in the relative proportions of individual PCE categories have been larger between 1960 and 1970 than the changes forecast from 1968 to 1985. Therefore, the use of the budgetary data for 1968 should not unduly bias the welfare losses forecast in this analysis.

MEASURES OF WELFARE LOSS

The problem of obtaining a monetary measure of the welfare loss imposed on a consumer by a change in prices has a long history in economics. In elementary textbooks, it is normally described in terms of changes in the area under demand curves. The argument runs as follows. Consider the demand curve for widgets shown in Exhibit IV-10(a). Suppose the retail market price is $5 per widget; at this price the consumer buys 4 widgets and spends a total of $20. However, this is less than the maximum amount of money which the consumer would be willing to pay to have 4 widgets. If widgets could only be purchased in single units, our demand curve implies that the consumer would pay up to $12 for his first widget. If he were subsequently offered a second widget, without being allowed to renegotiate the price of the first widget, he would pay up to

213

Exhibit IV-8 Changes in Consumption Patterns for Selected Categories		
Type of Product	1960 %	1970 %
Food for home consumption	19.1	16.2
Purchased meals, beverages	5.0	4.7
Tobacco	2.1	1.8
Women's and children's clothing	4.6	4.6
Men's and boy's clothing	2.5	2.5
Jewelry and watches	1.7	0.7
Shoes	1.4	1.3
Clothing services	1.0	0.9
Toilet articles, preparations	0.9	1.0
Personal care services	0.7	0.7
Housing (owner and tenant occupied)	14.2	14.7
Furniture, bedding	1.4	1.3
Household appliances	1.5	1.5
Other house furnishings	2.7	3.1
Household supplies	1.1	0.9
Household utilities	4.2	3.9
Drugs, supplies, equipment	1.4	1.4
Medical care services	4.5	6.2
Personal business	4.6	5.7
User-oriented transportation	12.2	11.7
automobile purchase	5.4	5.1
gasoline, oil	3.8	3.6
Local transportion	0.6	0.4
Intercity transportation	0.4	0.5
Radio, television	1.0	1.5
Toys, sporting goods	1.4	1.7
Books, magazines, newspapers	1.1	1.2
Paid admissions	0.5	0.4
Private education, research	1.1	1.7
Religious, welfare activities	1.5	1.4
Foreigh travel and other travel, net	0.7	0.8

SOURCE: The Conference Board

214

Exhibit IV-9

Composition of Personal Consumption Expenditures Categories for 1968 and 1985

Personal Consumption Expenditure Categories	1968	1985
1. Food purchased for off-premise consumption	16.3	15.0
2. Purchased meals and beverages	4.6	4.8
3. Tobacco Products	1.8	1.6
4. Shoes and other footwear	1.3	1.2
5. Women's & children's clothing and accessories	4.7	4.0
6. Men's & boy's clothing & accessories	2.6	2.2
7. Cleaning, dyeing, pressing, storage and repair of garments	0.8	0.3
8. Jewelry and watches	0.7	0.7
9. Toilet articles and preparations	1.0	1.1
10. Barbershops, beauty parlors and baths	0.7	0.6
11. Owner-occupied non-farm dwellings	9.5	10.2
12. Tenant-occupied non-farm dwellings	4.0	4.6
13. Furniture including mattresses & bedsprings	1.4	1.4
14. Kitchen and other household appliances	1.5	1.4
15. China, glassware, tableware & utensils	0.6	0.6
16. Other durable house furnishings	1.5	1.5
17. Semi-durable house furnishings	1.0	0.8
18. Cleaning and polishing preparations	0.9	1.4
19. Stationery and writing supplies	0.4	0.3
20. Electricity	1.5	1.8
21. Gas	0.9	0.7
22. Water and other sanitary services	0.4	0.3
23. Telephone and telegraph	1.5	1.8
24. Drug preparations and sundries	1.1	1.3
25. Ophthalmic products & orthopedic appliances	0.3	0.3
26. Privately controlled hospitals & sanitariums	2.4	2.7

Personal Consumption Expenditure Categories	1968	1985
27. New cars and net purchases of used cars	6.1	5.6
28. Tires, tubes, accessories and parts	0.8	0.8
29. Auto repair, greasing, washing, parking, storage and rental	1.4	1.5
30. Gasoline and oil	3.6	3.0
31. Street & electric railway and local bus	0.3	0.1
32. Taxicabs	0.1	-
33. Railway (commutation)	-	-
34. Railway (excluding commutation)	-	-
35. Intercity bus	0.1	0.4
36. Airline	0.4	-
37. Other	0.5	0.4
38. Books and maps	0.6	0.6
39. Magazines, newspapers and sheet music	0.9	0.8
40. Nondurable toys and sports supplies	0.7	1.0
41. Wheel goods, durable toys, sports equipment, boats and pleasure craft	1.5	1.7
42. Radio and television receivers, records, and musical instruments	0.2	0.1
43. Radio and television repair	0.2	0.3
44. Flowers and seeds, and potted plants	0.4	0.1
45. Admissions to spectator amusements	0.2	0.3
46. Clubs & fraternal organizations, except insurance	0.8	0.8
47. Higher education	0.4	0.6
48. Elementary and secondary schools	0.4	0.4
49. Other	83.0%	81.1%
Total		

$10 for the second widget. Similarly, he would pay up to $7
if he had to buy the third widget separately, and he would pay
$5 for the fourth widget. If he had bought the 4 widgets in
this manner, the consumer would have paid $34 for them. This
is $14 more than he would pay if he could buy them at a uni-
form price; this differential is approximately equal to the
triangular area under the demand curve, down to the $5 price
line--i.e., the shaded area in Exhibit IV-10(a). It represents
the consumer's net benefit from being able to purchase widgets
at $5 per unit. If the price of widgets now rose to $7 (e.g.,
because of an upward shift in the supply curve due to pollution control requirements), the consumer's net benefit would
be smaller--the consumer is less well off when purchasing
widgets at $7 than at $5. The welfare loss--the decrease in
the consumer's surplus--is shown by the cross-hatched area in
Exhibit IV-10(b). It is common to take this area as an ap-
proximate monetary measure of the welfare loss due to the
price change.

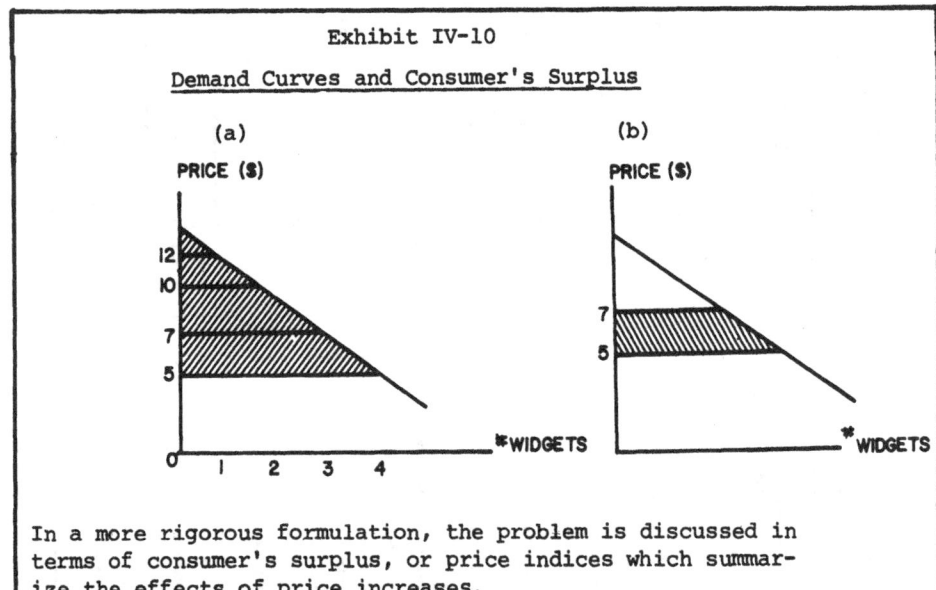

Exhibit IV-10

Demand Curves and Consumer's Surplus

In a more rigorous formulation, the problem is discussed in
terms of consumer's surplus, or price indices which summarize the effects of price increases.

It should be noted that welfare losses are well defined
only for individual consumers.

The basic postulate underlying the concept of consumer's
surplus is that the individual consumer pruchases commodities
as though he were maximizing a utility function, U(x), subject
to the constraint that his total expenditure equals his income,

216

i.e., $\sum_i P_i x_i = Y^*$, where P=price, x=purchases, Y=income. The
solution to this problem is a set of ordinary demand functions
relating the quantities of each good purchased to the prices
of all goods and the consumer's income:

$$X = h(P,Y)$$

Substituting the demand function back into the utility function (U) yields the indirect utility function (V):

$$U = U[h(P,Y)] = V(P,Y) \qquad \ldots(1)$$

which shows how the consumer's utility (welfare) varies with
prices and his income <u>when he optimally adjusts his purchase
to take account of changes in prices or income.</u> It can be
shown that when prices fall or income rises, the consumer is
better off (i.e., V increases), as one would expect. The welfare loss from a price change may be defined as the extra income the consumer would need to have when faced with one set
of prices in order to be as well off as he would be when facing a different set of prices, at possibly a different income
level.** For a convenience of exposition we shall assume that
the consumer's income (Y) is the same in both situations.
Assume that price change from P^0 to P^1. The welfare change
may be represented by the dollar sum, C, which satisfies the
following condition.

$$V[P^1, Y^0 + C] = V[P^0, Y^0] \qquad \ldots(2)$$

C is the amount by which the consumer's income must be augmented
in order that he should enjoy the same welfare when faced with
P^1 as he did when he had income Y^0 and faced prices P^0. For
future reference, we shall denote this (initial) welfare level
by $U^0 = V(P^0, Y^0)$.

The same concept may be derived in an alternative way, by
means of the so-called expenditure function. The minimum income
which a consumer must have and spend when faced with prices P,
in order to experience a welfare level U is a function of P and U.
This may be written as:

$$Y = M(P,U) \qquad \ldots(3)$$

The M-function may be thought of as being derived by inverting the
V-function to obtain Y in the terms of P and U (cf. equation (1)).
In terms of the M-function, the minimum income which the consumer
needs when faced with P^1 in order to experience the same level of

---

*In what follows X and P are vectors, i.e., $X=(x_1 \ldots x_n)$ and
likewise $P=(P_1 \ldots P_n)$; Y (the consumer's income) is, of course,
a scalar.

**We assume implicitly that the consumer's tastes are the same
in the two cases.

welfare as when prices were P^0 and his income was Y^0 is $M(P^1,U^0)$. Hence the amount by which the consumer's income must be augmented is, simply,

$$C = M(P^1,U^0)-Y^0 \qquad \ldots \text{ (4)}$$

Both (2) and (4) provide a precise definition of the welfare measure.

We may wish to express C as a proportion of the consumer's income, Y^0. We write this as:

$$\frac{C}{Y^0} = \frac{M(P^1,U^0)}{Y^0} -1 \qquad \ldots \text{ (5)}$$

The ratio on the right-hand side of (5) is of special interest, because it is the true cost-of-living price index for the change in prices from P^0 to P^1. For any cost of living index can be formally defined as the ratio of the costs of attaining the same standard of living under two different price regimes. The cost of achieving a level of welfare equal to U^0 when prices are P^0 is precisely Y^0; the cost of achieving the same level of welfare when prices are P^1 is $M(P^1,U^0)$ by definition. Hence, the cost of living index is merely the ratio of two terms whose difference is defined as the compensating variation measure of welfare loss.

C is not the only measure which could be used for assessing the welfare loss due to a price change. For example, we could derive a measure based on the idea of modifying the consumer's income so that, if faced with prices P^0 he would enjoy the same level of welfare as he does with prices P^1 and income Y^0. This concept, the converse of C, is the equivalent (E) variation measure of welfare loss. It can be formally defined as:

$$V(P^0,Y^0-E) = V(P^1,Y^0) = U^1 \qquad \ldots \text{ (6)}$$

or, equivalently, as

$$E = Y^0 - M(P^0,U^1). \qquad \ldots \text{ (7)}$$

As one might expect, E leads to a numerically different measure of the welfare loss imposed by a price change than C, except in special circumstances. Moreover, it can be shown that in the circumstances encountered in this study, C is likely to be a larger number than E.

It might be considered disconcerting that there is more than one--in fact, many more--potential measure of the welfare loss due to a price change. In truth, however, it is neither more nor less disconcerting than the fact that there is more than one way to define a cost of living index. A cost of living index can be defined with respect to any of a large number of base standards of living. So, too, with welfare measures of price changes. This does not mean that there is anarchy; each base standard of living and, correspondingly, each cost of living index or each measure of welfare loss has a different meaning and implies a different value judgement in assessing the effects of a price change. Thus, the C measure of welfare loss implies that the appropriate standard for comparison is the level of welfare before the price change, while the E measure implies that the appropriate standard for comparison is the consumer's welfare loss subsequent to the price change. Although we are calculating both measures of welfare loss, it is our subjective judgement that the C measure is to be. preferred in the present case.

CALCULATING THE WELFARE LOSSES

In order to calculate the welfare losses which result from the price increases caused by the requirements of P.L. 92-500, we have estimated both the compensating and equivalent variations according to formulas (2) and (6) in the previous section of this chapter. These calculations were made for 1977, 1980, and 1985 for the SEAS 1 scenario, which assumes timely compliance with the BPT and BAT requirements of the Act. In addition, C and E were calculated for 1985 for the SEAS 4 scenario, which considered the effects of BPT only.

C and E were calculated according to the following formulae:

$$C = P^O X^O \left[\left(\frac{P'X'}{P^O X'} \cdot \frac{P'X^O}{P^O X^O} \right)^{\frac{1}{2}} -1 \right] \quad \ldots \quad (1)$$

$$E = P'X' \left[1 - \left(\frac{P^O X'}{P'X'} \cdot \frac{P^O X^O}{P'X^O} \right)^{\frac{1}{2}} \right] \quad \ldots \quad (2)$$

where   P' = prices with population abatement for 1977, 1980, and 1985 (row vector)
         X' = quantities purchased with pollution abatement for 1977, 1980, and 1985 (column vector)
         X^O = quantities purchased without pollution abatement (base case) for 1977, 1980, and 1985 (column vector)
         P^O = prices without pollution abatement (base case) for 1977, 1980, and 1985 (row vector)'

The calculations of the compensating and equivalent variations through formulae (1) and (2) yield approximations of the welfare losses experienced as a result of the price increases; they represent exact measures only under some restrictive conditions.

In the model, prices were calculated in terms of index numbers, and quantities were measured in units of "one 1971 dollar's worth." For example, assume that in 1971 the price of widgets was $5, and 8 widgets were consumed. The quantity then in terms of the "new" units is 40 units of widgets. If the quantity of widgets increases from 8 widgets to 10, this may be expressed in terms of our new units as 50 units of widgets. Base case quantities were forecast for each year through the INFORUM model. The changes in the quan-

tities purchased which result from the price increases were estimated on the basis of price elasticities for each PCE category.
The price elasticity estimates were derived by Houthaker and Taylor.*

The Results

The welfare losses resulting from price increases in the SEAS
1 scenario** for 1977, 1980, and 1985 are described in Exhibits
IV-11, IV-12, and IV-13 respectively. These tables present estimates of the welfare loss measured by the compensating variation
(C) per family for each of the 10 income groups. This is then expressed as a fraction of the mean "total" income for each income
group. The total welfare loss by income group, and the percentage
distribution of the burden are also described.

As the equivalent variation represents a less relevant measure
of welfare loss for the purposes of this analysis, and because its
values were so close to the compensating variation values, only the
per family equivalent loss was presented in these tables.

The average welfare loss, or loss in real income will be approximately $142 for 1977, $280 for 1980, and $430 for 1985. This
welfare loss is 0.57% of the average "total" income per family for
1977; 1.03% for 1980; and 1.43% for 1985. These estimates are
slightly understated, as calculations of welfare losses were only
performed for 49 PCE categories, accounting from some 80% of the
total budget, but including all categories where major increases
were expected. The burdens caused by industrial price increases
show substantial growth over the period considered, both in absolute terms and relative to income. It should be noted, however,
that under the assumptions of the SEAS 1 scenario, the expenditures
necessary for compliance with the requirements of the Act will have
been incurred by 1985--therefore, price increases should not occur
beyond 1985. Further, even in 1985, the burden is less than 1.5%
of the total income for the average family.

The per family compensating variation (welfare loss) grows
as income increases; this is to be expected as total personal
consumption expenditures also grow with income. The per family
equivalent variation (welfare loss) is in all cases slightly
smaller than the corresponding compensating variation.

The welfare loss resulting from the price increases relative
to total income is the highest for the lowest income group: 1.35%
for 1977, 2.44% for 1980, and 3.34% for 1985. Although the price

---

*H.S. Houthaker and L.D. Taylor, Consumer Demand in the United
States, Cambridge, Mass.: Harvard University Press, 1970.

**Note that the SEAS 1 scenario assumes timely compliance with
the BPT and BAT requirements of the Act.

Exhibit IV-11

The 1977 Burdens of Industrial Price Increases under SEAS 1*

Income Group	Per Family C**	Per Family E**	C as % of Income	Total C in $Millions**	% Distribution C
1	$ 43.6	$ 43.0	1.35%	$ 348.4	3.8%
2	$ 44.5	$ 44.2	0.57%	$ 202.5	2.2%
3	$ 47.1	$ 46.9	0.43%	$ 299.6	3.2%
4	$ 68.3	$ 67.8	0.47%	$ 532.1	5.8%
5	$156.8	$155.5	0.86%	$1,456.7	15.8%
6	$179.4	$177.8	0.75%	$2,854.3	31.0%
7	$268.5	$265.9	0.84%	$1,358.6	14.8%
8	$268.8	$266.4	0.58%	$1,483.8	16.1%
9	$272.9	$271.3	0.27%	$ 532.2	5.8%
10	$272.9	$271.3	0.07%	$ 141.9	1.5%
Total				$9,210.1	100.0%

*SEAS 1 assumes Best Practicable and Best Available Technology (BPT and BAT) expenditures will be made on schedule.
**In $1973. C&E are Compensating and Equivalent measures of welfare loss as defined in the text.

Exhibit IV-12

The 1980 Burdens of Industrial Price Increases under SEAS 1*

Income Group	Per Family C**	Per Family E**	C as % of Income	Total C in $Millions**	% Distribution C
1	$ 86.4	$ 84.2	2.44%	$ 751.6	3.8%
2	$ 87.8	$ 87.1	1.02%	$ 436.4	2.2%
3	$ 92.9	$ 92.1	0.78%	$ 646.6	3.2%
4	$134.9	$133.3	0.84%	$ 1,149.3	5.8%
5	$310.1	$305.5	1.55%	$ 3,147.5	15.8%
6	$354.8	$349.5	1.35%	$ 6,170.0	31.0%
7	$531.4	$523.3	1.51%	$ 2,944.0	14.8%
8	$532.1	$524.5	1.04%	$ 3,208.6	16.1%
9	$539.4	$534.5	0.48%	$ 1,148.9	5.8%
10	$539.4	$534.5	0.13%	$ 307.5	1.5%
Total				$19,910.4	100.0%

*SEAS 1 assumes Best Practicable and Best Available Technologies (BPT and BAT) expenditures will be made on schedule.
**In $1973. C&E are the Compensating and Equivalent measures of welfare loss as defined in the text.

Exhibit IV-13

The 1985 Burdens of Industrial Price Increases under SEAS 1*

Income Group	Per Family C**	Per Family E**	C as % of Income	Total C in $Millions**	% Distribu- tion C
1	$131.5	$128.5	3.34%	$ 1,226.9	3.7%
2	$133.6	$132.4	1.40%	$ 709.4	2.2%
3	$140.9	$140.0	1.06%	$ 1,046.9	3.2%
4	$205.5	$203.8	1.16%	$ 1,944.0	5.9%
5	$475.8	$469.1	2.14%	$ 5,162.4	15.8%
6	$545.3	$537.4	1.87%	$10,131.7	30.9%
7	$819.3	$806.9	2.10%	$ 4,850.3	14.8%
8	$820.3	$808.4	1.45%	$ 5,290.9	16.2%
9	$829.5	$822.2	0.67%	$ 1,891.3	5.8%
10	$829.5	$822.2	0.18%	$ 506.0	1.5%
Total				$32,759.8	100.0%

*SEAS 1 assumes Best Practicable and Best Available Technologies (BPT and BAT) expenditures will be made on schedule.
**In $1973. C&E are the Compensating and Equivalent measures of welfare loss as defined in the text.

increases forecast for the necessities, which make up the largest proportions of the budget for these income groups, were not particularly large, the very high consumption relative to income accounts for these proportions. Welfare losses for income groups 2, 3, and 4 are only slightly higher than the welfare losses for the lowest income group. As income rises faster than the burdens for these groups, their welfare losses decline relative to income. The modest increase in the welfare losses experienced by these income groups may be explained in terms of their budget structure and propensities to consume. The consumption of necessities, such as food, clothing, and housing, continues to represent a significant proportion of the budget for these income groups; with the price increases being relatively modest for these PCE categories. Further, the propensity to consume declines with income.

Welfare losses jump for income groups 5, 6, and 7, both in absolute terms and relative to income. The welfare loss expressed as a percent of income is higher for each of these groups than the national average for 1977, 1980, and 1985. As their propensity to consume continues to decline, their relatively high burdens must be caused by their consumption patterns. In addition to categories #20 and #22, electricity and water, higher than average price increases are expected for #8, jewelry and watches, #10, barbershops and beauty parlors, #16, durable house furnishings, #25, opthalmic and orthopedic products, for example. In general, the categories experiencing higher than average price increases have income elasticities greater than one, their consumption increases as income grows. For income groups 8, 9, and 10, the welfare losses increase

222

only slightly; thus, the compensating variation, expressed as a percent of income, decreases sharply. Consumption, as a proportion of total income is decreasing for these groups; therefore, the ratio of C/Y also drops.

The total burden for each income group was obtained by multiplying the per family C estimate by the appropriate number of families. The distribution of the burdens between families does not change over time. This distribution is presented in Exhibit IV-14. Our assumption that consumption patterns do not change over time is partially responsible for this. Although the magnitude of the price increases caused by pollution control increase over time, their structure (ratio to the national average price increase) remain constant.

For analyzing the incidence of industrial price increases, an alternative scenario--SEAS 4-- was considered. In this scenario the effects of BPT alone are analyzed, with the necessary investments completed by 1980. The results for 1985 are presented in Exhibit IV-15.

The average welfare loss (compensating variation) is $231, which is less than the $280 estimate for the SEAS 1 scenario for 1980 (Exhibit IV-12). SEAS 1 also assumes BPT by 1980. The estimates of welfare losses by family and total welfare losses by income group and the percentage distribution of welfare losses by income group are similar for these two cases. Price increases for the SEAS 4 scenario have been described in the previous section of this chapter; their structure is similar to the structure of price increases for the SEAS 1 case. Therefore, the distribution of the burdens by income groups may be expected to remain constant. Per family burdens relative to income are lower for the SEAS 4 case; the burdens are approximately equal for 1980 SEAS 1 and 1985 SEAS 4. Since income is expected to grow between 1980 and 1985, relative burdens decrease.

Exhibit IV-14

Lorenz Curve: 1985 Distribution of the Burdens Caused by
Price Increases under SEAS 1

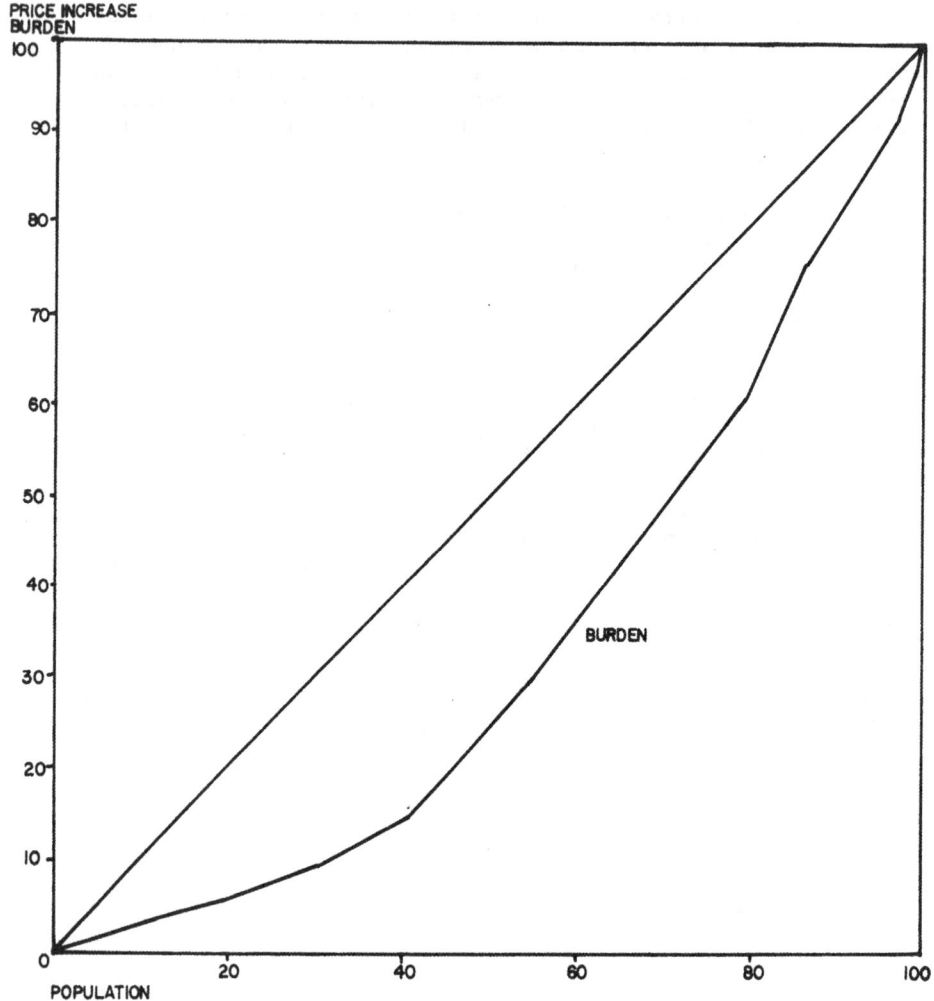

Exhibit IV-15

The 1985 Burdens of Industrial Price Increases under SEAS 4*

Income Group	Per Family C**	Per Family E**	C as % of Income	Total C in $Millions**	% Distribution C
1	$ 83.1	$ 82.0	2.11%	$ 775.3	3.8%
2	$ 84.6	$ 84.2	0.89%	$ 449.2	2.2%
3	$ 89.5	$ 89.0	0.68%	$ 665.0	3.2%
4	$130.3	$129.5	0.73%	$ 1,232.6	6.0%
5	$300.2	$297.7	1.35%	$ 3,257.2	15.8%
6	$343.8	$340.8	1.18%	$ 6,387.8	30.9%
7	$515.7	$510.9	1.32%	$ 3,052.9	14.8%
8	$516.4	$511.8	0.91%	$ 3,330.8	16.1%
9	$523.0	$521.9	0.42%	$ 1,192.4	5.8%
10	$523.0	$521.9	0.11%	$ 319.0	1.5%
Total				$17,612.8	100.0%

*SEAS 4 assumes that only Best Practicable Technology (BPT) expenditures will be made on schedule.
**In $1973. C&E are Compensating and Equivalent measures of welfare loss as defined in the text.

V. Conclusions

What then will be the total burden of water pollution control costs on different income groups? Will the burden be large? Will the distribution of costs among different socio-economic groups be equitable? Before analyzing the total burden, let us review quickly the findings in Chapter III on the public costs resulting from P.L. 92-500 and the findings in Chapter IV on the industrial costs resulting from the Act.

Public Costs

Local governments generally use the property tax and user charges to finance the bulk of treatment plant construction and maintenance, and to retire bonds. The federal government, which supplies up to 75% of the funds for many publicly owned treatment works, relies mainly on the personal income tax. From public finance studies, we know how each of these and other methods bear on households of differing circumstances. Thus, by determining the particular mechanisms used to finance treatment facilities, the distribution of costs can be inferred.

In order to compare the costs of public treatment works for different socioeconomic groups, we have assumed that the public construction programs represented by Categories I, II, and IVB or Categories I-V of EPA's 1974 Needs Survey would be completed by 1983. Categories I, II, and IVB cover treatment plant and interceptor sewer construction that the EPA felt that public bodies needed to undertake. Categories I-V add collector sewers, infiltration-inflow control, sewer rehabilitation, and combined sewer overflow. To forecast the methods of financing to be used (charges and/or property taxes, for example), we analyzed fiscal trends in the past and conducted a survey of local governments.

For the average* family, the expected tax increases to pay for the construction costs represented by Needs Survey Categories

---

*The average family is defined here as one in the median income group.

227

I, II, and IVB are approximately $51* by 1977, $58 by 1980, and
$63 by 1985. Construction costs for Categories I-V of the Needs
Survey imply tax increases of $133 in 1985 for the average family.
Since this represents less than 1% of the average total family
income,** no major distributional impacts are anticipated. The
distribution of the incremental tax burden is roughly similar to
the federal tax burden, which is generally described as progressive. In other words, the poor will not be paying a disproportionately large share of the costs.

Industrial Costs

    To meet pollution control requirements, industries will incur
costs which are likely to be reflected in the prices they charge.
Those who buy the products of the firms, in effect, pay the costs
of the new treatment facilities. By knowing how purchases of
products are distributed between different kinds of households,
we can estimate how much each will pay for the required treatment
facilities. For example, a relatively small increase in food
prices has a relatively large effect on the lower income groups,
because these groups spend a large proportion of their income on
food. The welfare losses caused by price increases have been calculated for each income group on the basis of their consumption
patterns. For the average family, these are $157 in 1977, $310 in
1980, and $476 in 1985. These burdens are substantially higher
than the burdens imposed by public facilities. However, the
burdens imposed by industrial costs still represent a small fraction (less than 2.5%) of total income. The price increases resulting from the Act are thus not expected to alter the distribution of income to any significant extent. It should be noted,
however, that the distribution of the price increase burden is
different from the distribution of the tax burden: the middle income groups are particularly hard hit.

Combined Costs

    The total burden for the average family is $208 in 1977,
$368 in 1980, and $539 in 1985, representing approximately 1.14%,
1.84%, and 2.43% of total family income. The distribution of income is not appreciably altered by pollution control costs, because these costs are small relative to total income. The total
burdens increase with income; in 1985 the total burden for the
lowest income group is $158, and more than ten times that amount
for the highest. Pollution control costs decline as a percent of
income, however. For the lowest and highest income groups, these

---

*NOTE: All costs are in 1973 constant dollars.

**Total family income includes earnings, unreported income, fringe
benefits, retained corporate profits, and imputed rent on owner
occupied housing. See Chapter II.

proportions are approximately 4.0% and 0.4% respectively. This might be considered a disproportionate burden for the poor. However, the incidence of pollution control costs is roughly comparable to the distribution of the federal tax burden. The lowest two income groups pay 6.1% of the federal tax burden, and 6.2% of the pollution control burden.

The magnitude of the total burden relative to income varies over time as well as over income groups. Exhibits V-1 to V-3 present the estimates of the burdens caused by tax and price increases for 1977, 1980, and 1985.*

In 1977, for the lowest income group (#1), the average tax increase necessary to pay for publicly-owned treatment facilities is $16.60 per year per family. The price increase burden is substantially higher; it is $43.60. The total burden sums to $60.20, which represents 1.86% of the average total family income in this group. In absolute terms, the burdens increase with income; for the fifth (median) income group, the total burden is $207.90. For the highest income group, it is $1254.60. Relative to income, however, the burdens decrease; for the median group, 1.14% of the total family income is "lost" through tax and price increases. For the highest income group, only 0.33% is "lost."

Pollution control expenditures in both the public and private sector increase over time; the burdens on individual families grow as well. This is described in the bar graphs of Exhibits V-4 to V-6. By 1980, the total burden of the average family in the lowest income group increases to $108.60, and by 1985 it grows to $158.30. These represent respectively 3.06% and 4.02% of the average total family income. In other words, pollution control burdens grow not only in dollars but also as a percent of income. For the median income group, the burdens are $368.30 for 1980, or 1.84% of their income, and $539.40 for 1985, or 2.43% of their income.

The impacts of the costs incurred by the public and private sectors may also be compared in Exhibits V-1 to V-3. In general the price increase burden is greater than the burden imposed by additional taxes. This is true for all but the highest income group. Exhibit V-7 describes the percentage of the total burden caused by price increases for selected years for the 10 income groups. This proportion increases over time, indicating that private sector costs grow faster than public sector expenditures.

---

*These calculations are based on the assumption that the requirements of the Act will be satisfied through private sector investments in both Best Practicable Technology (BPT) and Best Available Technology (BAT) and public sector investment in the amounts described in the 1974 Needs Survey categories I, II, and IVB. For a more detailed description of our assumptions see Chapters III and IV.

Exhibit V-1

1977 Water Pollution Control Cost Burdens By Income Group

Income Group	Annual Costs Per Family (rounded $1973)			Tax Burden as a % of Income	Price Increase Burden as a % of Income	Total Burden as a % of Income
	Tax Burden*	Price Increase Burden**	Total Burden			
1	$ 16.60	$ 43.60	$ 60.20	0.51%	1.35%	1.86%
2	$ 29.00	$ 44.50	$ 73.50	0.37%	0.57%	0.94%
3	$ 37.20	$ 47.10	$ 84.30	0.34%	0.43%	0.77%
4	$ 47.60	$ 68.30	$ 115.90	0.33%	0.47%	0.80%
5	$ 51.10	$156.80	$ 207.90	0.28%	0.86%	1.14%
6	$ 67.60	$179.40	$ 247.00	0.28%	0.75%	1.03%
7	$110.60	$268.50	$ 379.10	0.35%	0.84%	1.19%
8	$ 70.70	$268.80	$ 339.50	0.15%	0.58%	0.78%
9	$186.40	$272.90	$ 459.30	0.19%	0.27%	0.46%
10	$981.70	$272.90	$1254.60	0.26%	0.07%	0.33%

*Tax burden figures are based on construction to meet Needs Categories I, II, IVB.

**Price increase burden is based on a zero baseline and on the assumption that Best Practicable and Best Available Technology (BPT and BAT) expenditures will be made on schedule.

Exhibit V-2

1980 Water Pollution Control Cost Burdens By Income Group

| Income Group | Annual Costs Per Family in Rounded $1973 | | | Tax Burden as a % of Income | Price Increase Burden as a % of Income | Total Burden as a % of Income |
	Tax Burden*	Price Increase Burden**	Total Burden			
1	$ 22.20	$ 86.40	$ 108.60	0.62%	2.44%	3.06%
2	$ 35.20	$ 87.80	$ 123.00	0.41%	1.02%	1.43%
3	$ 43.60	$ 92.90	$ 136.50	0.36%	0.78%	1.14%
4	$ 54.60	$134.90	$ 189.50	0.34%	0.84%	1.18%
5	$ 58.20	$310.10	$ 368.30	0.29%	1.55%	1.84%
6	$ 75.40	$354.80	$ 430.20	0.28%	1.35%	1.63%
7	$120.90	$531.40	$ 652.30	0.34%	1.51%	1.85%
8	$ 78.80	$532.10	$ 610.90	0.16%	1.04%	1.20%
9	$207.80	$539.40	$ 747.20	0.19%	0.48%	0.67%
10	$999.80	$539.40	$1539.20	0.24%	0.13%	0.37%

*Tax burden figures are based on construction to meet Needs Categories I, II, and IVB.

**Price increase burden is based on a zero baseline and on the assumption that Best Practicable and Best Available Technology (BPT and BAT) expenditures will be made on schedule.

231

Exhibit V-3

1985 Water Pollution Control Cost Burdens By Income Group

Income Group	Annual Costs Per Family in Rounded $1973			Tax Burden as a % of Income	Price Increase Burden as a % of Income	Total Burden as a % of Income
	Tax Burden*	Price Increase Burden**	Total Burden			
1	$ 26.80	$131.50	$ 158.30	0.68%	3.34%	4.02%
2	$ 40.30	$133.60	$ 173.90	0.42%	1.40%	1.82%
3	$ 48.50	$140.90	$ 189.40	0.37%	1.06%	1.43%
4	$ 59.90	$202.50	$ 262.40	0.34%	1.16%	1.50%
5	$ 63.60	$475.80	$ 539.40	0.29%	2.14%	2.43%
6	$ 81.10	$545.30	$ 626.40	0.28%	1.87%	2.15%
7	$ 127.90	$819.30	$ 947.20	0.33%	2.10%	2.43%
8	$· 87.70	$820.30	$ 908.00	0.15%	1.45%	1.60%
9	$ 213.80	$829.50	$1043.40	0.17%	0.67%	0.84%
10	$1013.40	$829.50	$1842.90	0.22%	0.18%	0.40%

*Tax burden figures are based on construction to meet Needs Categories I, II, and IVB.

**Price increase burden is based on a zero baseline and on the assumption that Best Practicable and Best Available Technology (BPT and BAT) expenditures will be made on schedule.

Exhibit V-4

Graph of 1977 Water Pollution Control Costs by Income Group

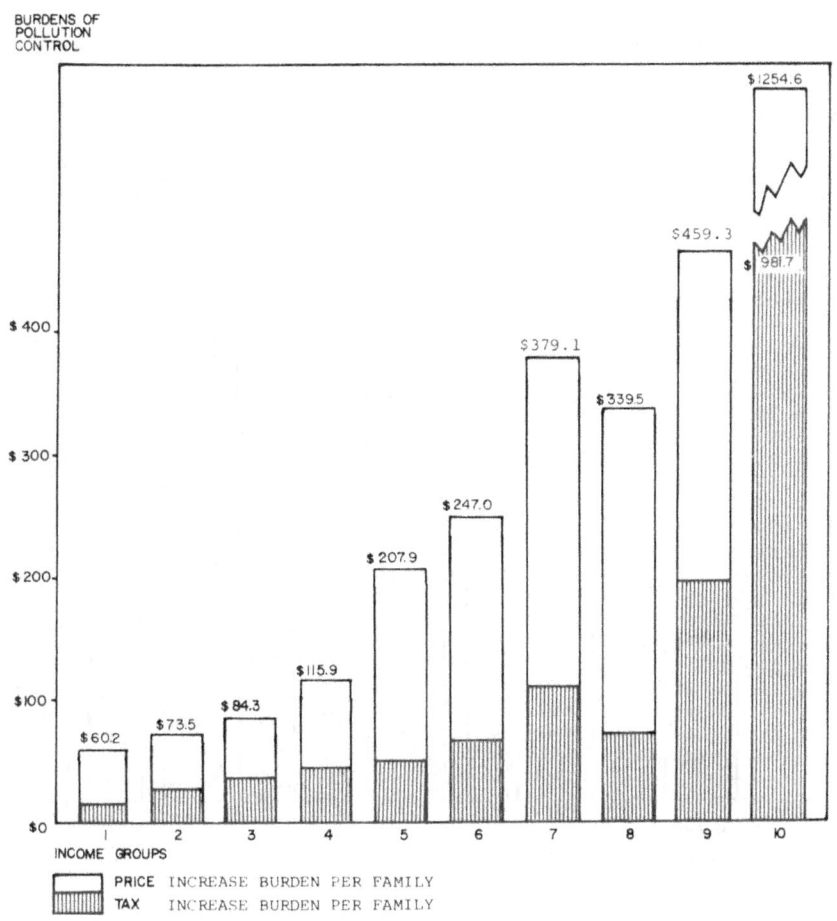

BURDENS OF
POLLUTION
CONTROL

$1254.6

$459.3

$ 98l.7

$400

$379.1

$339.5

$300

$247.0

$207.9

$200

$115.9

$100

$84.3

$73.5

$60.2

$0

INCOME GROUPS

1   2   3   4   5   6   7   8   9   10

PRICE INCREASE BURDEN PER FAMILY

TAX INCREASE BURDEN PER FAMILY

233

Exhibit V-5

Graph of 1980 Water Pollution Control Costs by Income Group

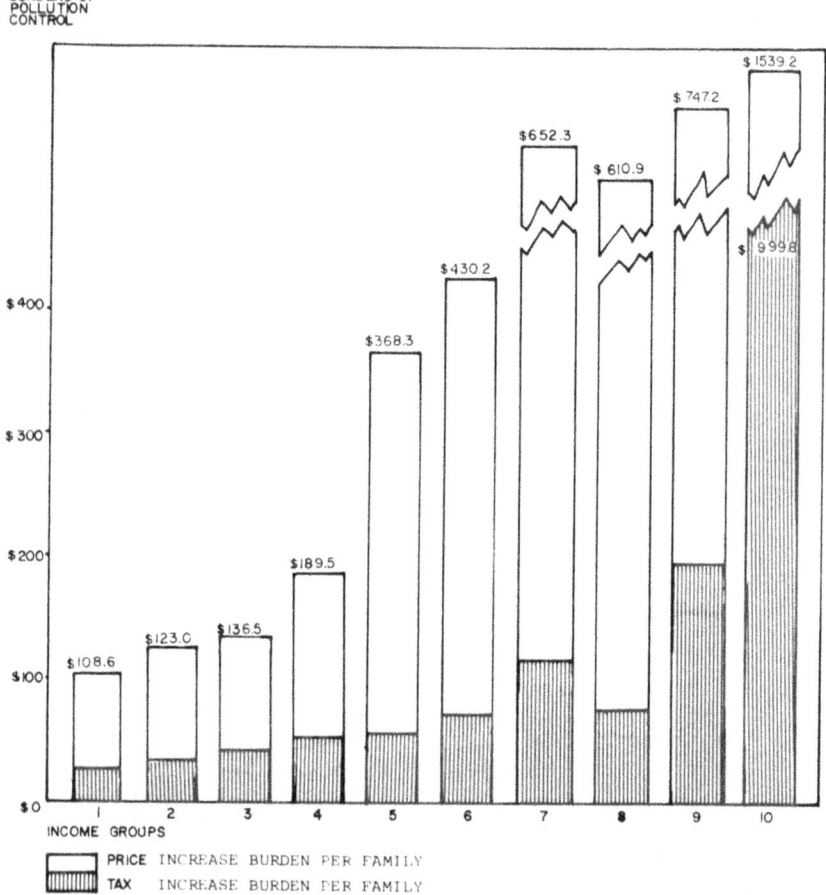

234

Exhibit V-6

Graph of 1985 Water Pollution Control Costs by Income Group

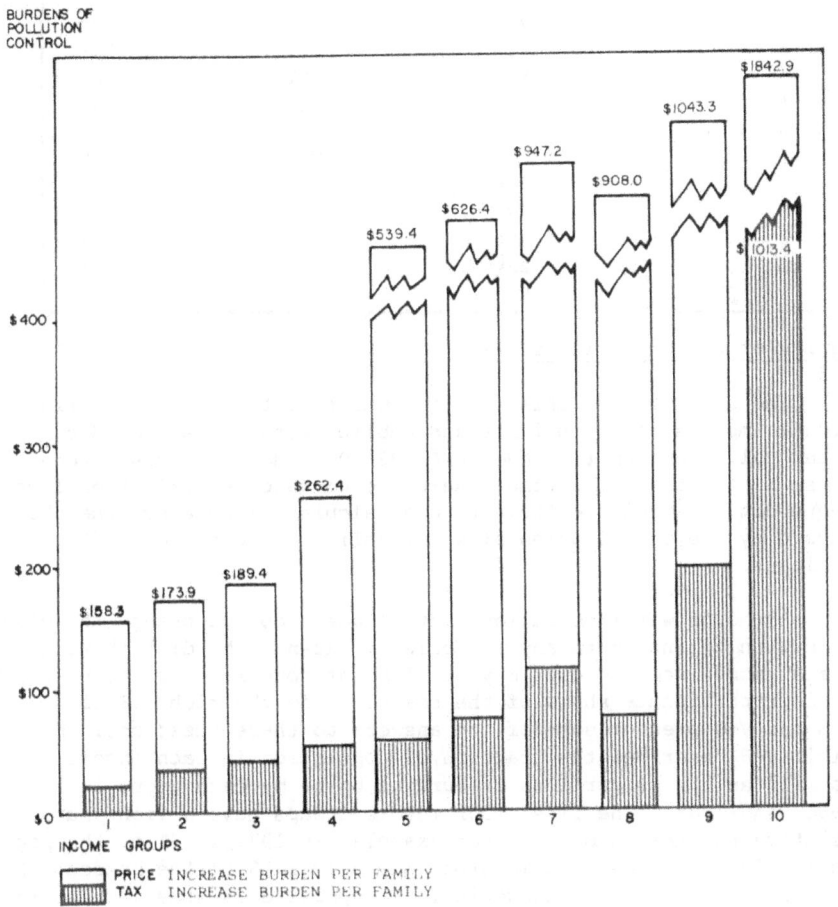

BURDENS OF
POLLUTION
CONTROL

PRICE INCREASE BURDEN PER FAMILY
TAX INCREASE BURDEN PER FAMILY

INCOME GROUPS

235

Indeed, the tax burden represents a progressively smaller proportion of total income over time; the opposite is true for the
price increase burden.

Exhibit V-7

Price Increase Burden as a Percent of the Total Burden
Caused By Water Pollution Control Costs

Income Group	1977	1980	1985
1	72%	80%	83%
2	61%	71%	77%
3	56%	76%	74%
4	59%	71%	77%
5	75%	84%	88%
6	73%	82%	87%
7	71%	81%	86%
8	79%	87%	90%
9	58%	72%	80%
10	22%	35%	45%

Equity of the Cost Distribution

The objective of this research was to estimate the distributional impacts of the private and public costs necessitated by
water pollution control under P.L. 92-500. Because this burden
is small relative to income, there are no major distributional consequences. A Gini coefficient* was calculated; however, the changes
caused by the distribution of water pollution costs were not significant.

Although water pollution control costs do not change the distribution of income to any appreciable extent, the distribution of
costs among income groups may still be of concern. Do the poor pay
a disproportionate share of the costs, or do the rich? Exhibits
V-8 and V-9 present some of the answers to these questions. Exhibit V-8 describes the fraction of population in each income group,
as well as the proportions of burdens borne by each group in 1977,
1980, and 1985. The first four income groups have a light burden
relative to their numbers. For example, in 1977 12.3% of the population in the lowest income group pay only 3.6% of the burden. For
the six higher income categories, the opposite is true. These relationships hold over time, although some minor changes do occur
in the burden distribution.

*A Gini coefficient is a measure of income concentration (and
thus of equity among income groups). It is defined in Chapter II.

Exhibit V-8

The Distribution of Water Pollution Control Costs

Burdens for 1977, 1980, and 1985

Income Group	Percent of Population in Group	% of Total Burden Borne by Each Income Group		
		1977	1980	1985
1	12.3%	3.6%	3.8%	3.8%
2	7.0%	2.5%	2.4%	· 2.4%
3	9.8%	4.0%	3.8%	3.6%
4	12.0%	6.7%	6.4%	6.2%
5	14.3%	14.3%	14.9%	15.1%
6	24.5%	29.2%	29.8%	30.1%
7	7.8%	14.3%	14.4%	14.5%
8	8.5%	13.9%	14.7%	15.1%
9	3.0%	6.8%	6.3%	6.1%
10	0.8%	4.8%	3.5%	2.9%
Total	100.0%	100.0%	100.0%	100.0%

Exhibit V-9

The Distribution of the Total Burdens of Water Pollution
Control Costs (1985) and the Distribution
of the Federal Tax Burden

Income Group	Percent of Population in Group	% of Total Burden Borne By Each Income Group		
		Pollution Control Burden, 1985	Federal Tax Burden	Federal Personal Income Tax
1	12.3%	3.8%	2.7%	0.8%
2	7.0%	2.4%	3.4%	1.2%
3	9.8%	3.6%	6.0%	3.9%
4	12.0%	6.2%	9.5%	7.2%
5	14.3%	15.1%	12.1%	10.1%
6	24.5%	30.1%	26.6%	26.3%
7	7.8%	14.5%	13.6%	14.7%
8	8.5%	15.1%	8.4%	10.3%
9	3.0%	6.1%	7.5%	10.5%
10	0.8%	2.9%	10.2%	15.0%
Total	100.0%	100.0%	100.0%	100.0%

The distribution of the tax and price increase burdens and the total burdens are also depicted in Exhibit V-10 for 1985. In this figure, the Lorenz curve presents the cumulative burden distribution, by income group, against the cumulative population distribution, by income group. Cumulative burdens in percentage terms are depicted along the vertical axis, population along the horizontal, with population in order of increasing income (lowest income groups to the left). If the burdens were distributed so that everybody paid the same amount, the resulting cumulative distribution would be the diagonal, which is the equivalent of a poll tax (that is, a uniform per capita or per family tax). If the richest fraction of the population were to pay the entire cost, the cumulative distribution would be represented by a right angle along the horizontal and right hand vertical axes. In other words, the more regressive the distribution of the burden, the closer the curve is to the diagonal. For particular income groups--say the lowest 20% of the income distribution--shallow slopes indicate that their burdens are below the national average, steep slopes indicate the opposite. In Exhibit V-10, line 1 describes the price increase burden; line 2, the total burden; and line 3, the tax burden. Lines 1 and 2 are very close together indicating that the distribution of the total burden is largely dominated by price increases. In all these curves, the lower half of the population pays less than the national average costs and the upper income half pays more.

The distribution of water pollution control cost burdens relative to the distribution of population may be described as progressive. However, this amounts to no more than saying that pollution control costs are not as regressive as a poll tax. This is not particularly meaningful, as economic capacity varies between the ten income groups. A more meaningful yardstick is the distribution of the federal tax burden, assuming that (in some sense) this represents the democratic preference with respect to the distribution of burdens. The federal tax burden is also progressive relative to population, the lower five income groups pay a smaller share of federal taxes than their numbers, and the upper income groups pay proportionately more. For the lowest income group, the pollution control burden is higher than the federal tax burden. For income groups 2, 3, 4, 9, and 10, it is higher, and for income categories 5, 6, 7, and 8, it is lower. The differences, however, are not substantial except for the highest income group, which pays 2.9% of the pollution control costs in contrast to 10.2% of federal taxes.

The distribution of the personal income tax is also presented for comparative purposes, as this tax is normally regarded as progressive. The Lorenz curve in Exhibit V-11 describes the cumulative distribution of the pollution control cost burden, the total federal

EXHIBIT V-10

Lorenz Curve: Cumulative Distribution of Water Pollution
Control Cost Burdens, 1985

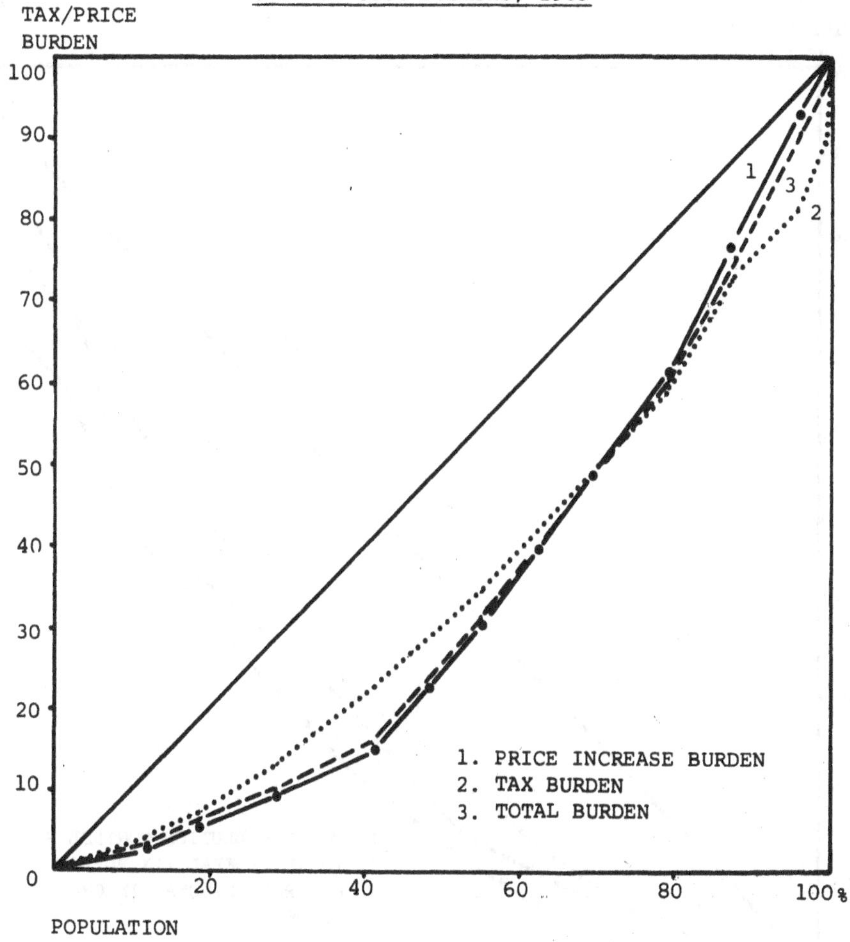

TAX/PRICE
BURDEN

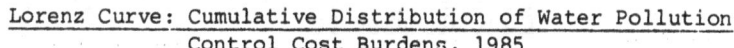

1. PRICE INCREASE BURDEN
2. TAX BURDEN
3. TOTAL BURDEN

POPULATION

NOTE: Tax burden figures are based on construction to meet
Needs Categories I, II, IV-B.
Price increase burden is based on a zero baseline and
on the assumption that Best Practicable and Best Available Technology (BPT and BAT) expenditures will be made
on schedule.

239

Exhibit V-11

Lorenz Curve: Cumulative Distribution of Water Pollution Control
Costs for 1985, Total Federal Taxes and the Federal
Personal Income Tax

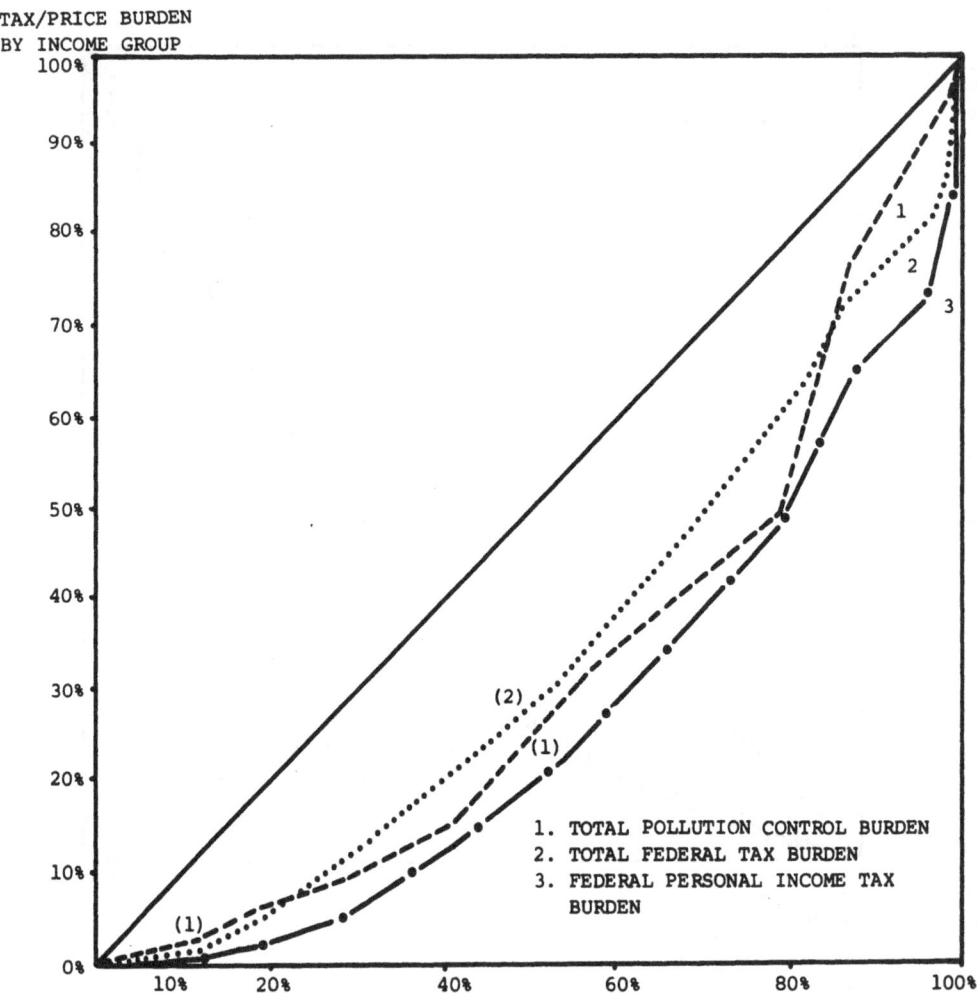

TAX/PRICE BURDEN
BY INCOME GROUP

POPULATION BY INCOME GROUP

1. TOTAL POLLUTION CONTROL BURDEN
2. TOTAL FEDERAL TAX BURDEN
3. FEDERAL PERSONAL INCOME TAX
   BURDEN

tax, and the federal personal income tax burden. The total federal tax burden and the water pollution control cost burdens are more regressive than federal personal income taxes for all income groups.

In addition to considering the distrubution of pollution control burdens for the entire U.S. population, certain subgroups were singled out for special attention. In particular, the burdens for groups with lower incomes, such as Blacks and selected age groups, were also considered. Regional differentials in the distribution of costs were estimated only for the costs of public treatment facilities. Such estimates were calculated through considering regional differences in costs and financing mechanisms, as discussed in Chapter III. Price increases were not calculated on a regional basis, not did we have data on consumption patterns by region. Because of the relative importance of the price increase burden, regional differentials in the total pollution control cost burden were not calculated.

Pollution control costs vary with income, as tax burdens and consumption patterns are both closely linked to family receipts. In view of this correlation, pollution control burdens for Blacks and selected age groups were calculated through the income link. Other factors, such as differences in the consumption patterns due to age and race, are ignored; therefore, the estimates described below are not exact.

For Blacks, the median income group is the third. Pollution control burdens for the average Black family total $84.30 in 1977, $136.50 in 1980, and $189.40 in 1985. These figures may be compared with similar estimates for the entire U.S. population, $207.90, $368.30, and $539.40 respectively. Burdens for the average Black family are lower, because their incomes are lower. No other factors were considered.

The distribution of pollution control burdens for Blacks and the entire U.S. population is described in Exhibit V-12. The distribution of the burdens is more regressive for Blacks than for the balance of the population; that is, lower income Blacks pay an unduly high share of the pollution control costs borne by Blacks. The reason for this is that the proportion of Blacks falling in the lower income groups is much higher than for the U.S. population as a whole.

Similar estimates were calculated for households in different age groups. Earnings are the lowest for households with heads under 25 or over 65 years of age, and earnings peak between 45 and 55 years of age. The total pollution control burden for the average household in the lowest and highest age groups for 1985 is $189.40. For the 45-55 age group, the median burden is $626.40. These estimates may be compared with the national figure, which is

241

Exhibit V-12

The Distribution of Water Pollution Control

Costs by Race and Income Group, 1985

Income Group	Blacks		U.S. Population	
	Percent of Population*	Percent of Costs Borne**	Percent of Population*	Percent of Costs Borne**
1	34.1%	18.0%	12.5%	3.8%
2	14.8%	8.5%	9.3%	2.4%
3	14.4%	9.0%	12.1%	3.6%
4	11.6%	10.0%	14.0%	6.2%
5	9.3%	16.8%	14.3%	15.1%
6	9.8%	20.5%	20.8%	30.1%
7	3.0%	9.5%	8.7%	14.5%
8	1.5%	4.6%	5.2%	15.1%
9	0.7%	2.4%	2.5%	6.1%
10	0.1%	0.6%	0.6%	2.9%
Total	100.0%	100.0%	100.0%	100.0%

*Percent of that population falling in the different income
groups

**Of the total water pollution control costs borne by that
population, the percentage borne by different income groups
within that population.

$539.40 per family. The distribution of the burden between income
groups is described for each age group in Exhibit V-13. The distribution of the burdens for the bottom half of the income distribution is the most regressive for the youngest and oldest age groups,
and the most progressive for the households with the age of head
between 45 and 55 years of age. These patterns reflect the differentials in earning capacity over the life cycle.

In sum, the costs of water pollution control, paid by households in the form of increased taxes and increased prices, represent
a relatively small proportion of total family income. The average
per family burden is approximately 1.14% of total family income in
1977, 1.84% in 1980, and 2.43% in 1985. Because of the relatively
small size of this burden, no major distributional impacts are anticipated.

Although total burdens increase in dollar amounts, these
costs do decline as a percent of income. For the lowest and highest income groups, these proportions are approximately 4.0% and
0.4% respectively. The incidence of pollution control costs is
nevertheless roughly comparable to the distribution of the federal
tax burden.

Exhibit V-13

The Distribution of Water Pollution Control Burdens by Age of Household Head

Of Costs Borne By Age Group, Percent Borne By Different Income Groups Within Age Group

Income Group	Age of Household Head					
	Under 25	25 - 34	35 - 44	45 - 54	55 - 64	Over 65
1	13.9%	4.0%	2.7%	2.4%	4.0%	18.1%
2	9.9%	3.1%	2.1%	1.8%	3.1%	12.8%
3	12.8%	5.8%	3.7%	3.2%	4.7%	8.6%
4	12.5%	9.7%	6.7%	5.6%	7.1%	7.3%
5	25.8%	22.7%	15.6%	13.1%	16.1%	14.6%
6	14.6%	28.9%	31.4%	29.2%	26.0%	15.6%
7 - 10	10.4%	26.7%	37.8%	44.6%	39.1%	22.8%
Total	100.0%	100.0%	100.0%	100.0%	100.0%	100.0%

Comparable figures on the population distribution by income group within each age group is provided in Exhibit II-11 in Chapter II.

Thus, the equity impacts of the Act appear small, and it does not appear likely that the poor will pay a disproportionate share of the costs.